TRICKS, GAMES AND PUZZLES WITH MATCHES

MAXEY BROOKE

Illustrated by Norman Dreyer

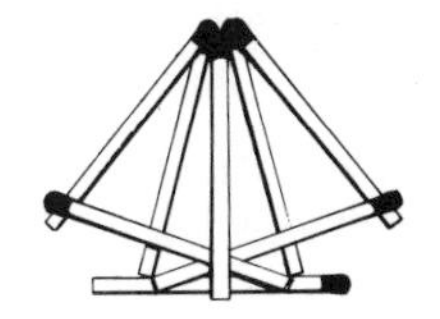

DOVER PUBLICATIONS, INC., NEW YORK

Published in Canada by General Publishing Company, Ltd., 30 Lesmill Road, Don Mills, Toronto, Ontario.
Published in the United Kingdom by Constable and Company, Ltd., 10 Orange Street, London WC 2.

Tricks, Games and Puzzles with Matches is a new work, first published in 1973 by Dover Publications, Inc.

International Standard Book Number: 0-486-20178-3
Library of Congress Catalog Card Number: 72-84816

Manufactured in the United States of America
Dover Publications, Inc.
180 Varick Street
New York, N.Y. 10014

BY WAY OF INTRODUCTION . . .

I was sitting in the PX during the war (World War II, that is), drinking beer with a friend. He laid a handful of matches on the table.

"Let's play a game," he said. "We'll take turns removing either one or two matches from the pile. The one who picks up the last match buys the beer."

I bought beer for the rest of the evening. It was my introduction to match tricks.

Because I am an inveterate collector of such things, I put my friend's game in my notebook. From time to time I added others.

Then one day, I found that I had almost a hundred of them. I asked Mr. Cirker of Dover Publications if he would be interested in publishing them. He said he would. And here they are.

Most of the tricks herein do not depend upon matches being matches. They can be done with toothpicks, soda straws, or even broomsticks. In fact, I recommend that parents give their children burned matches.

Some of the tricks are so simple that you will feel like kicking yourself when you see the solution. Others can only be described as elegant. The complete solution of at least one (No. 18) requires the knowledge of elementary number theory. Solutions to the puzzles begin on page 31.

I hope you enjoy working these puzzles as much as I enjoyed collecting them.

Good luck!

Notes

1. In various puzzles you may utilize square root as made from the matches:

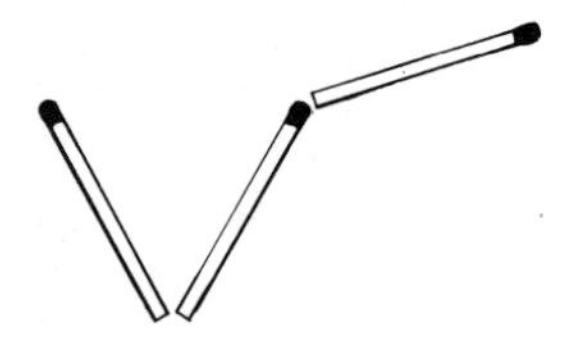

2. The bar above a Roman numeral denotes thousands, e.g.

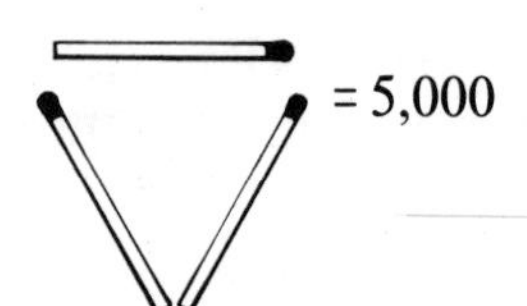

Matches, as we know them, have been in general use for less than 100 years. We don't know what genius first thought of using them for puzzles.

The earliest published puzzle I can find is in *Récréations Mathématiques* by the master puzzlesmith Édouard Lucas, written during the period 1884 to 1891. Among the three problems he gives is this one:

1. Remove three matches and leave three squares.

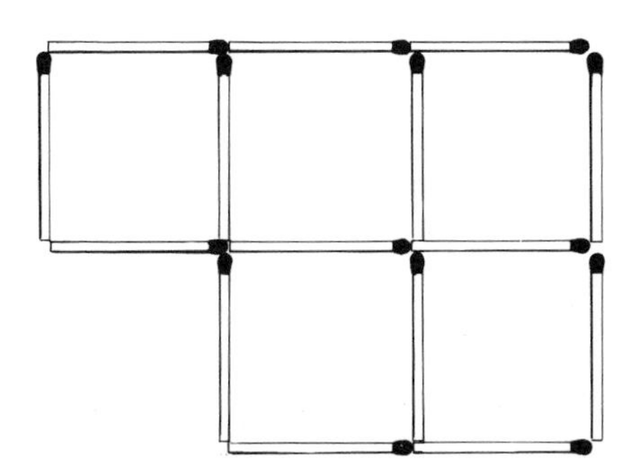

The other two puzzles from this early publication will be found elsewhere in this collection.

Here are six more, all starting with a three-by-three grid.

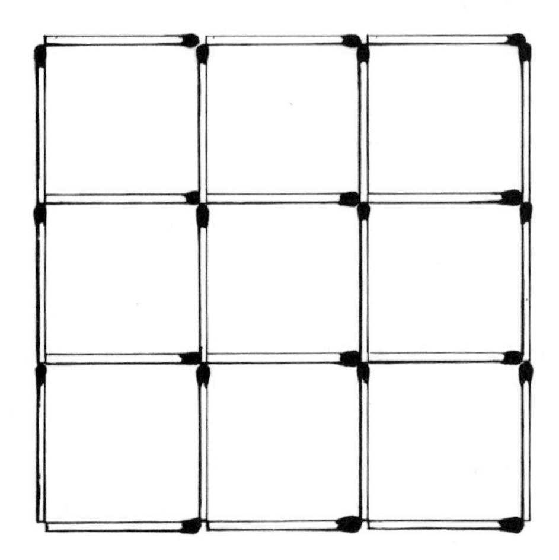

2. Remove four matches and leave five squares.

3. Remove six matches and leave five squares.

4. Remove six matches and leave three squares.

5. Remove eight matches and leave four squares.

6. Remove eight matches and leave three squares.

7. Remove eight matches and leave two squares.

You will note that it took 24 matches to make the grid pictured above. With the same 24 matches you can make one large square six matches per side; or two squares, both with three matches on each side; or three squares, with two matches per side.

Now, can you use 24 matches to construct the following?

8. Four squares.

9. Five squares.

10. Six squares.

11. Seven squares.

12. Eight squares.

13. Nine squares.

In addition to geometric problems, matches can be used to propose and solve problems in arithmetic.

14. Move one match and make this equation valid.

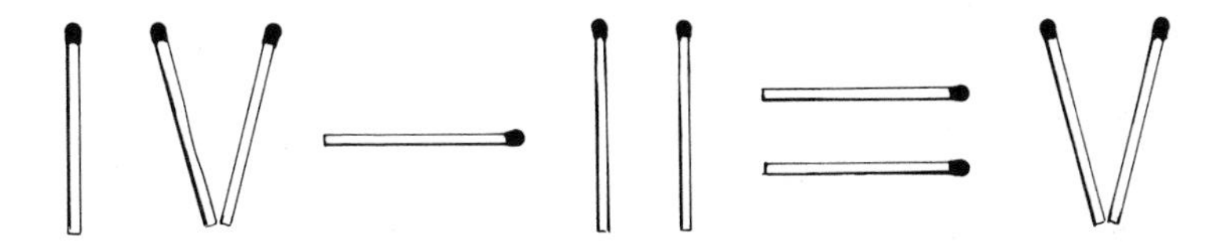

15. Remove two matches and make this equation valid.

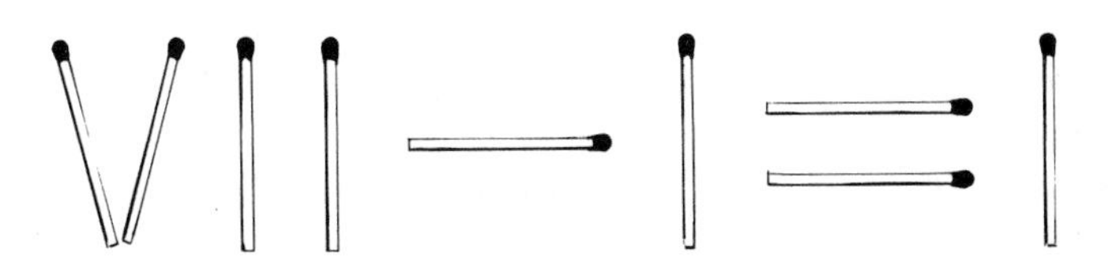

In addition to puzzles, there are match games, both solitary and competitive.

16. Here are eight matches.

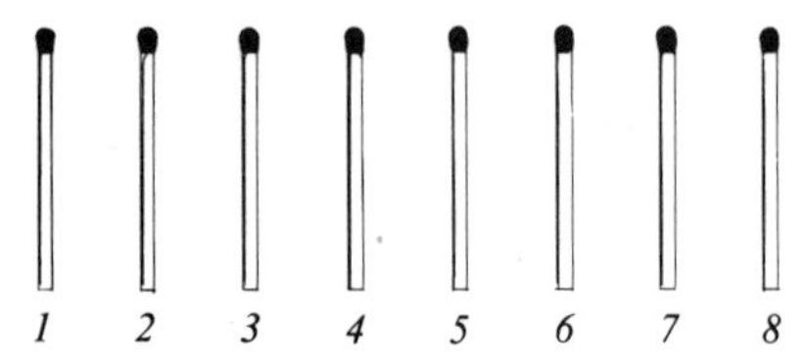

Can you make four crosses in four moves by picking up one match at a time and passing it over two others before laying it crosswise on the next match?

17. Now, start with a row of 15 matches. Jump three each time and end up with five piles of three matches each.

18. This one is called Parity Nim. Put 25 matches in a pile. Two players alternately take one, two, or three matches from the pile. When all the matches are gone, the player with the odd number wins. Can you devise a winning strategy for this game?

Matches provide the wherewithal for dozens of problems, tricks, and gimmicks. So here, without further comment, is a potpourri.

19. Here are 12 matches forming six equilateral triangles. Move four matches and leave three equilateral triangles.

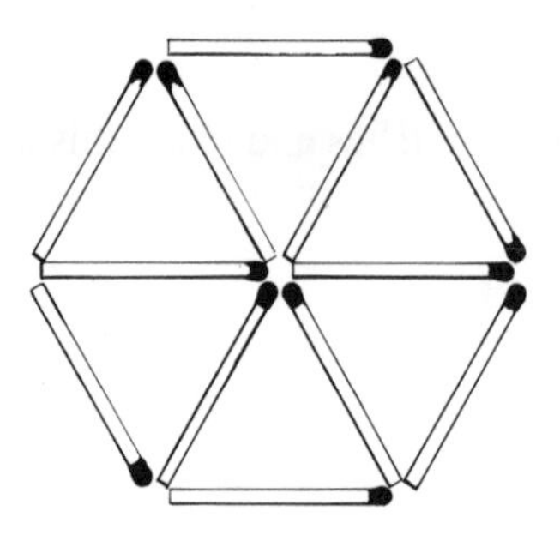

20. Move one match and make a square.

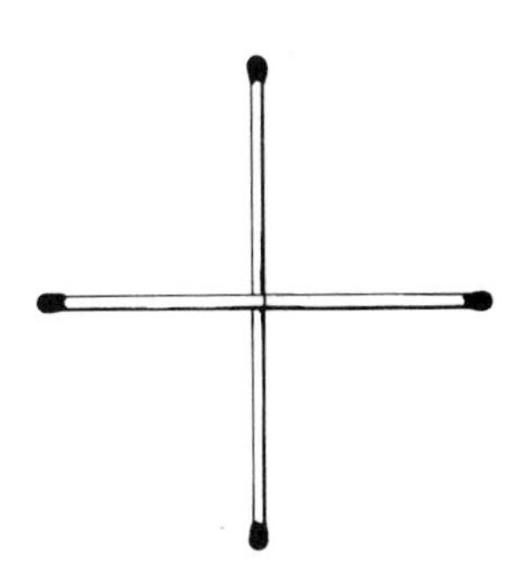

21. Move one match and make this equation valid.

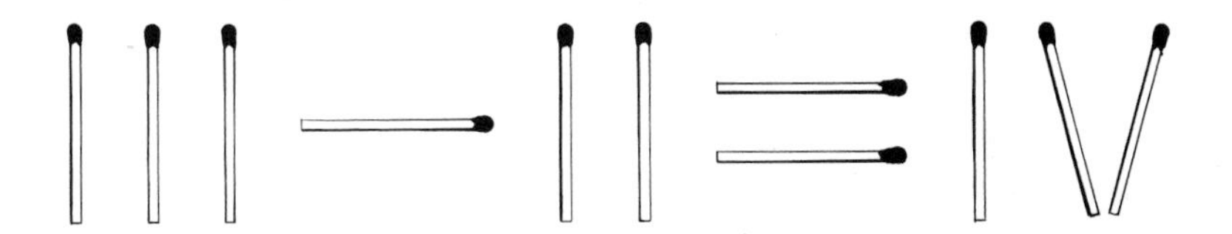

22. Challenge someone to break a match, using only the strength of the fingers, when it is held thus. (Although this looks easy, few can do it.)

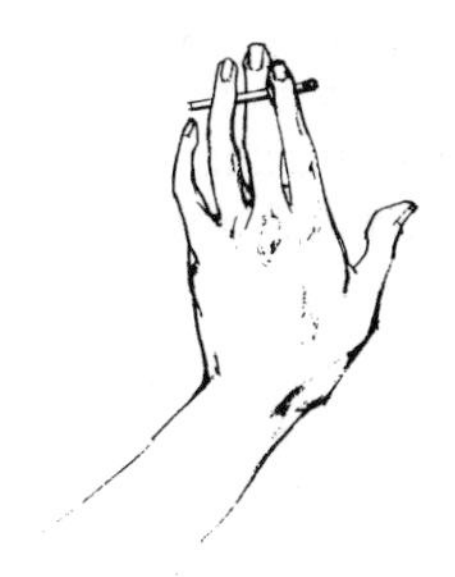

23. The big square is a farm. The smaller square is the homestead. The farmer wishes to retire and remain on the homestead. He also wants to divide the rest of the farm equally among his five sons. Can you do it for him?

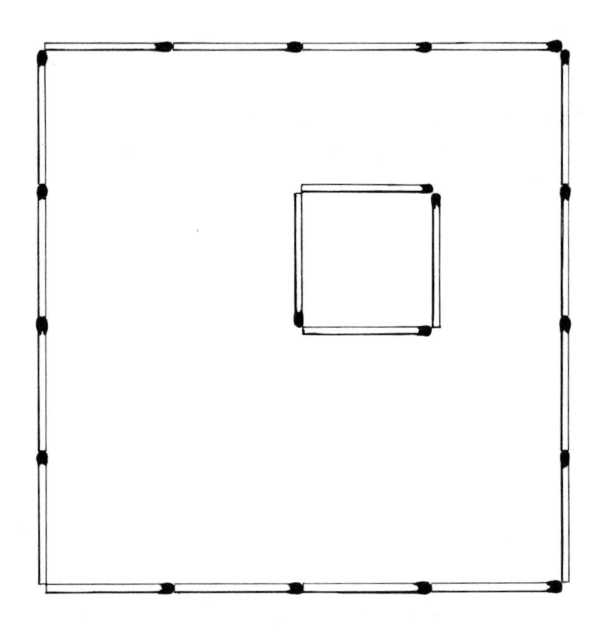

24. Here's an amusing trifle. Lay match *A* on the table and on top of it, match *B* at the angle shown. Match *C* goes on top of *B*, *D* on top of *C*, and so on. Match *A* should be close enough to *C* so that the head of *B* touches the table. If you press down on *F*, the head of *B* will rise from the table.

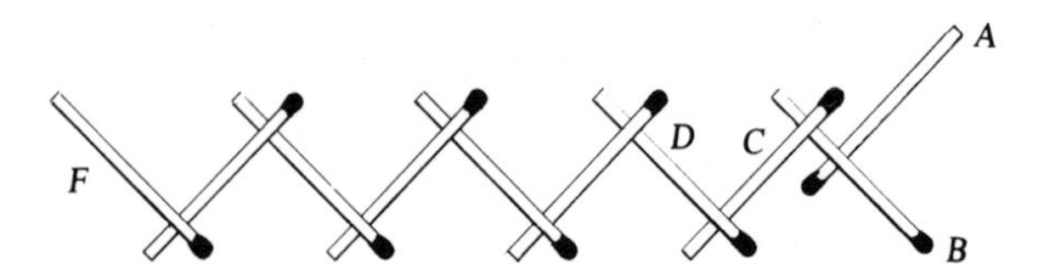

25. Move two of the matches to make a correct equation. There are at least three solutions.

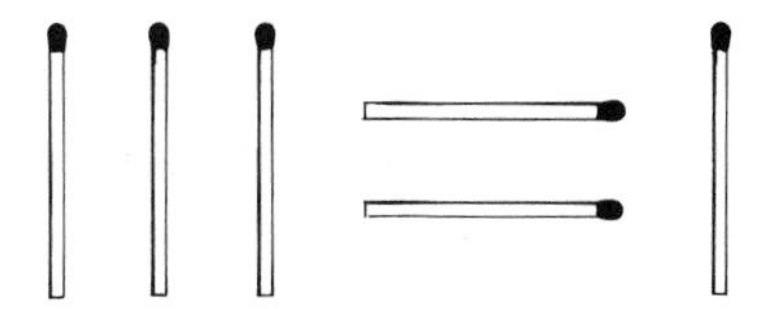

26. Balance about four matches across the mouth of an empty soft-drink bottle. About ten more matches can be laid across these four. Keep building until the entire edifice comes tumbling down. I once saw a picture in *Life* of more than 500 matches balanced thus. It also makes a good game. Players alternately balance matches. The one whose match causes the edifice to fall is the loser and must pay for the drinks.

27. Move one match and make the equation valid. There are two solutions.

28. Arrange eight matches to form two squares and four triangles.

29. Move three matches and end up with five triangles.

30. About 1890, that Prince of Puzzlesmiths, Édouard Lucas, invented the game called Pipoppipette, which became quite popular. It consisted of a board with 36 dots arranged in a square, and 60 bars. Players would place the bars so that each end was on one of the dots. When a player completed a square, he would place a colored counter in it. The player with the most squares would win.

As a schoolboy, I frequently played a variation of this game which we called Dots and Squares. I have since heard it called Square-it and the French Polytechnic Game. A square array of dots is used and the idea is for two players playing alternately to draw single lines horizontally or vertically between the dots. A player completing a square scores a point and the player with the most points after the array of dots is completely filled to form a grid wins the game.

It is only recently that this game has yielded to analysis. If you are interested, the winning strategy was published in *Recreational Mathematics Magazine* for August 1961.

A modification can be played with matches. I call it Continuous Polytechnic. Both players have 30 matches. They alternately lay down their matches one at a time. Each match played must touch the end of a match already played, either at a 90° or 180° angle. There can be no more than five matches in a horizontal or vertical row. The player completing a square scores a point. When all the matches have been played, the player with the most squares wins.

An alternate rule: When a player completes a square, he gets another play.

Try it. I think you will like it.

31. Express 100 using six matches.

32. Eighteen matches form a Solomon's seal which comprises eight triangles. Move two matches and reduce the number of triangles to six.

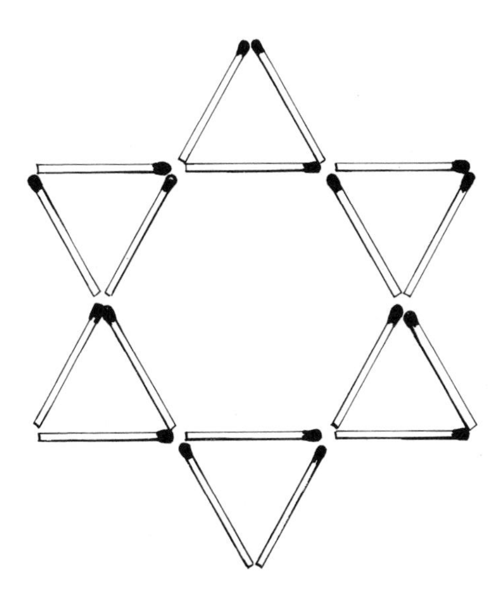

33. This is the first match game I ever devised. Hence I have a special affection for it, even though it's very simple.

1. Draw seven parallel lines less than a match length apart.
2. Each player has five matches.
3. Players alternately lay matches along lines with heads pointing toward themselves.
4. If two parallel matches are adjacent, players can put a match across them with the head pointing to his right.
5. Parallel matches count one point; crossed matches count two points.

The player with the most points wins.

34. Here are six matches. Add five more and make nine.

35. Lay eight matches parallel; then, on them, at right angles, lay six matches, so that each horizontal match touches each vertical match. How many parallelograms are formed?

36. This farmer has a large piece of land, and a homestead. How can he divide the land (apart from the homestead) equally among his six sons?

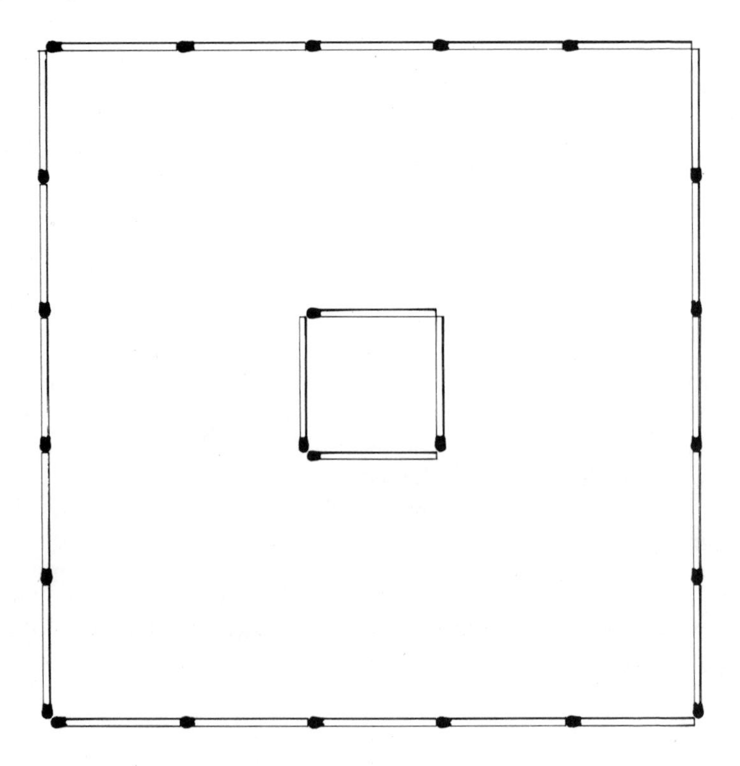

37. Could the farmer in No. 36 divide his property equally if he had eight sons?

38. Arrange six matches so that each touches the other five.

39. Move one match and make this equation valid.

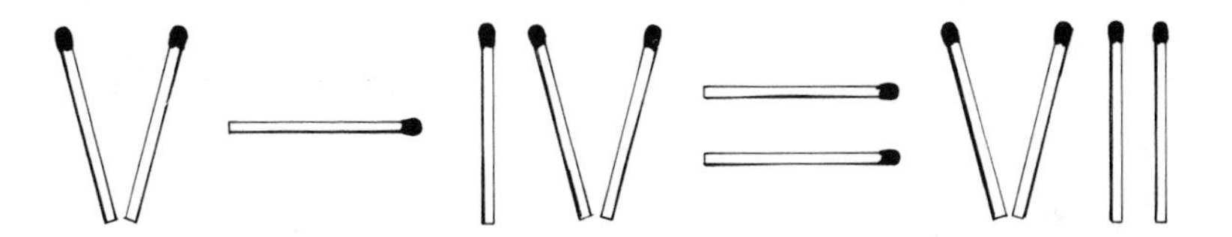

40. Make six squares with nine matches.

41.

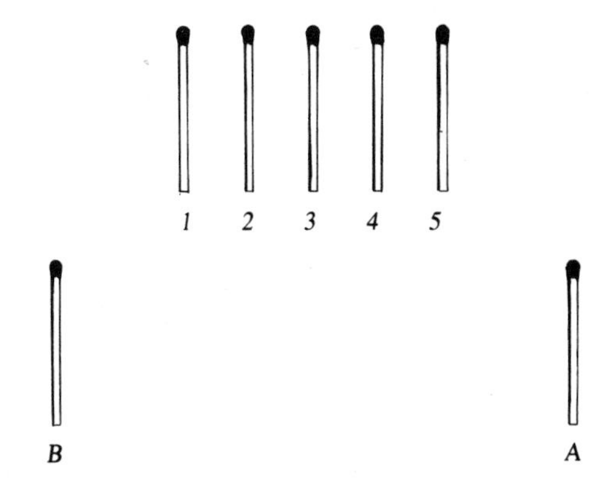

The five top matches represent five chickens. The two bottom ones represent two tramps. Entertain your friends by telling the following story while you do this trick.

Two hungry tramps are on a chicken scrounging expedition. *(Pick up match A in your right hand and match B in your left.)* In a certain henhouse, they find five chickens. First tramp A grabs a chicken *(pick up match 5 in your right hand);* tramp B grabs a second *(pick up match 1 in your left hand);* A gets another *(pick up match 4 in your right hand);* B still another *(pick up match 2 in your left hand);* and A the last one *(pick up match 3 in your right hand).*

They hear the farmer coming, so they quickly replace the chickens on the roost. *(Put the five matches back, starting with one from the left hand first; then from the right, the left, the right, and the left. You should now have two matches in your right fist and none in your left. Keep your hands closed throughout so your audience thinks you still have matches in both hands.)* The farmer looks into the henhouse, sees that the chickens are all right, and leaves. The tramps now grab the chickens again. *(Repeat, picking up the matches alternately, first with the right hand, then the left, the right, the left, and the right.)* They run off with the chickens and would have escaped if they hadn't gotten into a fight over their booty. It seems that tramp A ended up with four chickens and tramp B with only one. *(Open your hands. You now have five matches in your right and only two in your left.)* I've never been able to understand what happened.

42. Make this roman numeral equation read correctly without touching a match.

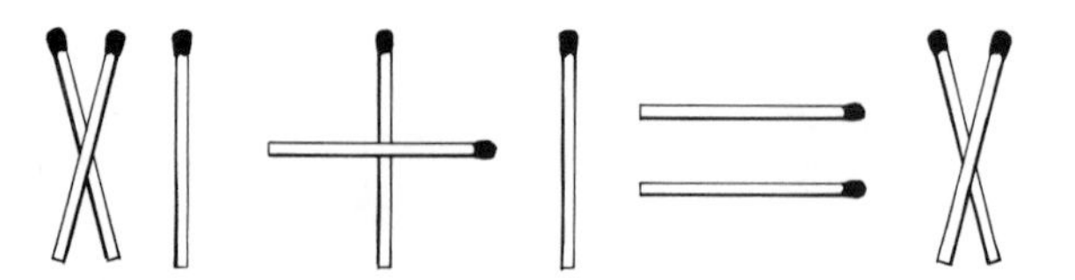

43. Remove two matches and leave three squares.

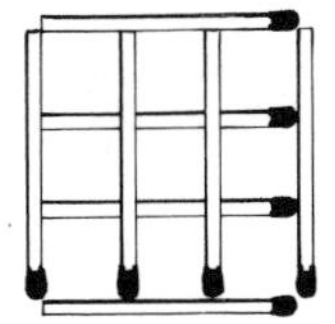

44. Problem: can you put four matches on the table and lay a quarter on them so that the quarter touches each match, but not the table; and the head of each match does not touch either the quarter or the table?

45. Move one match and make the equation valid.

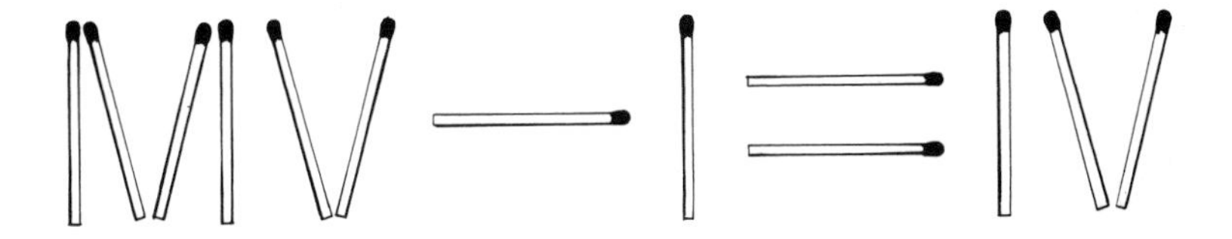

46. The Greek Temple. First, move two matches and make 11 squares. Then move four matches and make 15 squares.

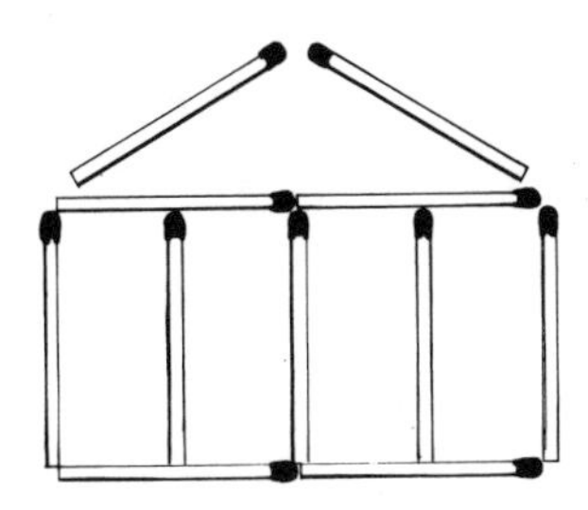

47. Make a square from six matches. Okay to cheat on this one!

48. Put a coin beneath a glass. Support a match between this glass and another. The trick: remove the coin without allowing the match to fall.

49. Change this fraction (1/6) to unity by adding only one more match.

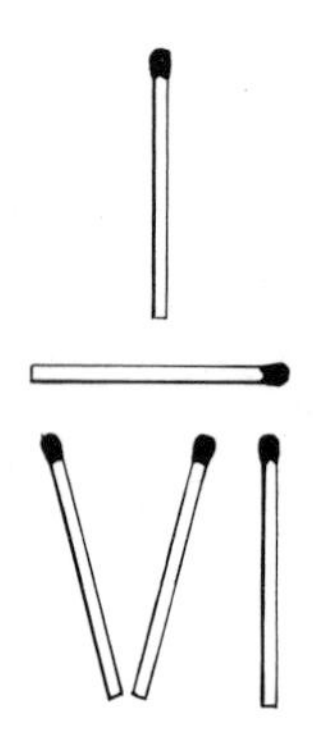

50. This is one way to make 10 using nine matches. Can you find another way?

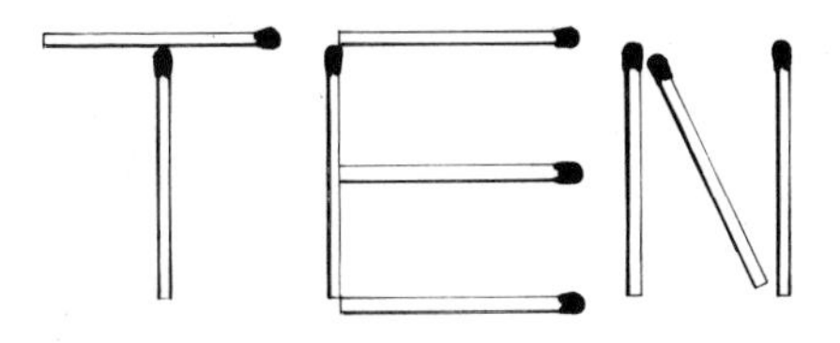

51. Move four matches and make three squares.

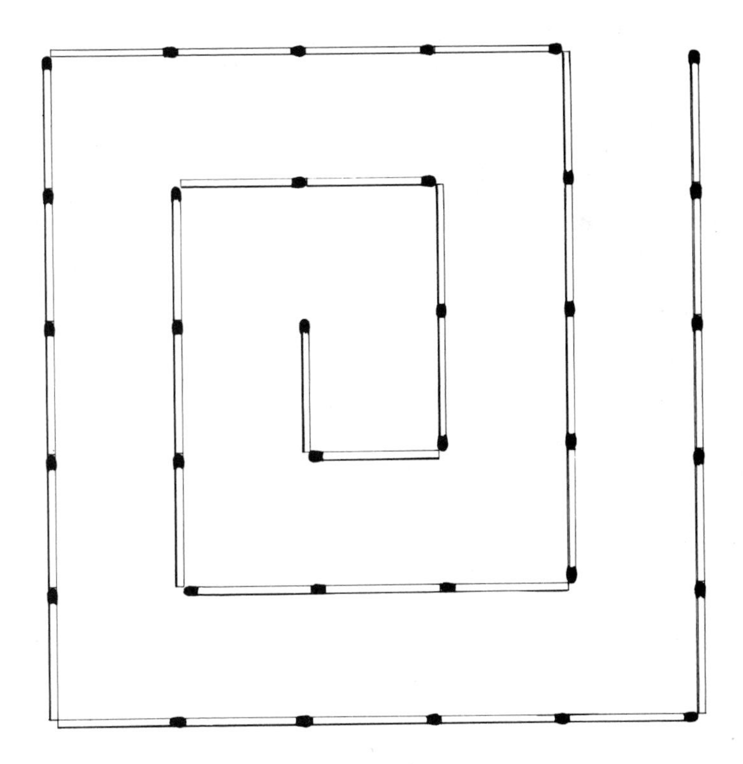

52. The area between the two squares is a moat inhabited by man-eating sharks. Our hero is marooned on the island in the center. His rescuers have only two matches. How can they build a bridge for him?

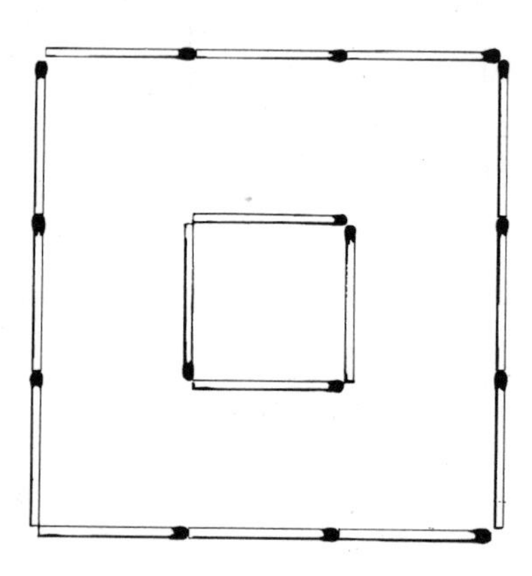

53. Hand six matches to someone and ask him to lay them on the table so that they will make "three-and-a-half-dozen."

54. Move two matches and make the equation valid.

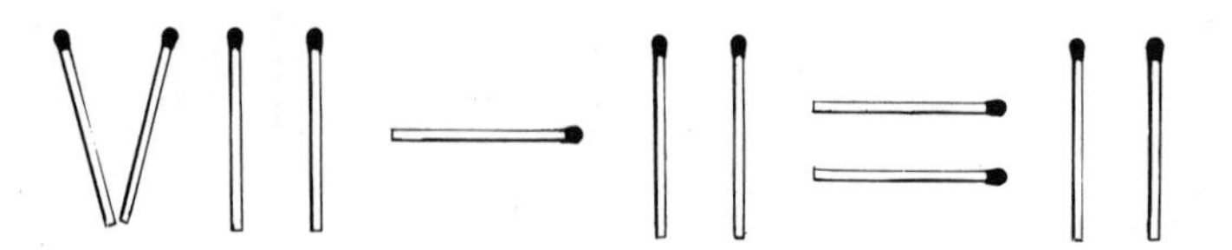

55. To the 16 matches that form this figure, add eight more, so as to divide the figure into four parts of equal size and shape.

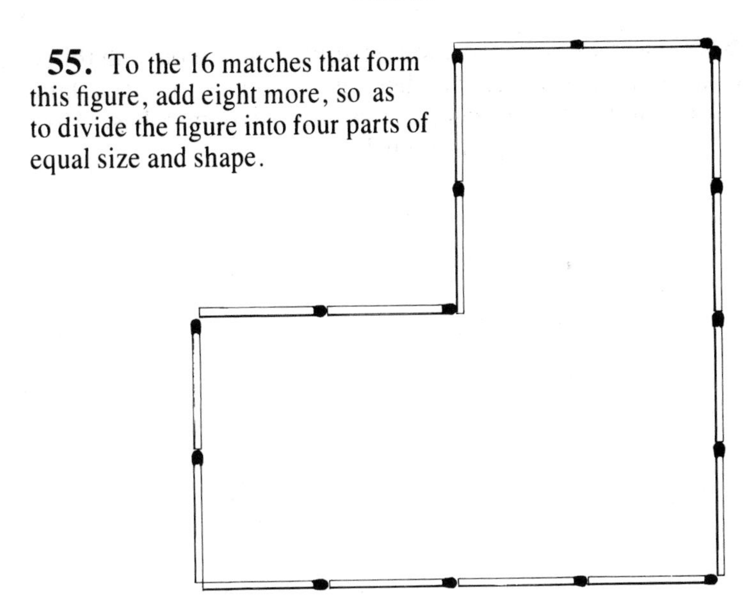

56. Make a slit in the end of a match and sharpen another to a chisel point. Push the chisel into the slit and prop up the two matches thus fastened together by resting another match *(A)* lightly against the junction. The trick: lift these three matches with a fourth match.

57. Move two matches to make four squares.

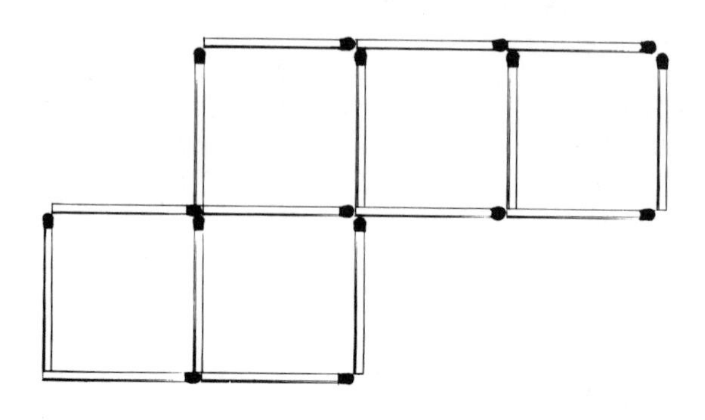

58. Move one match and make the equation valid.

59. Can you take six matches, break two in half, and arrange them to enclose three equal squares?

60. Make four squares with 16 matches.

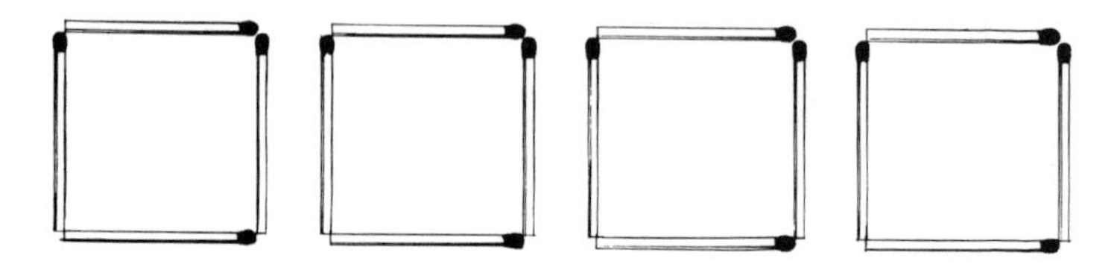

Now remove four matches and shift three to get "what matches are made of."

61. These 13 matches represent a sheep pen divided into six equal compartments. But suppose you have only 12 matches. How can you build a sheep pen that still has six equal compartments?

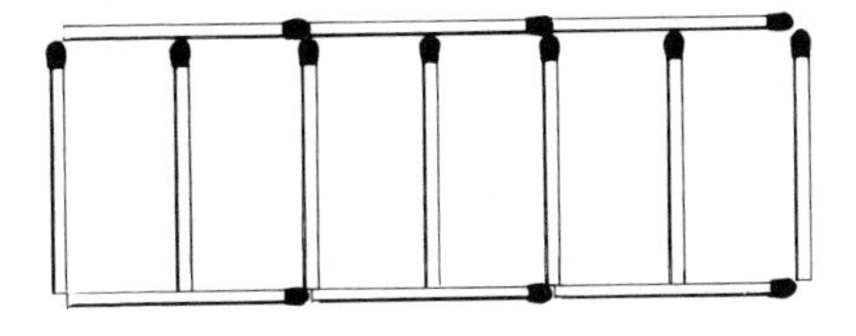

62. Move one match and make this equation valid.

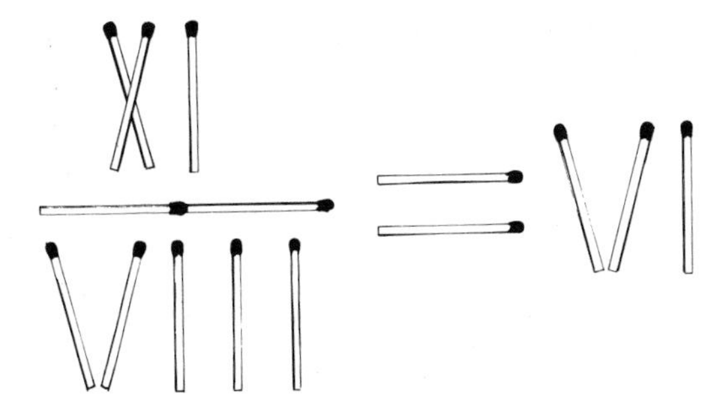

63. Here are 18 matches arranged to enclose two four-sided spaces, one twice as large as the other. Can you use 18 matches to enclose two five-sided spaces; one three times as large as the other?

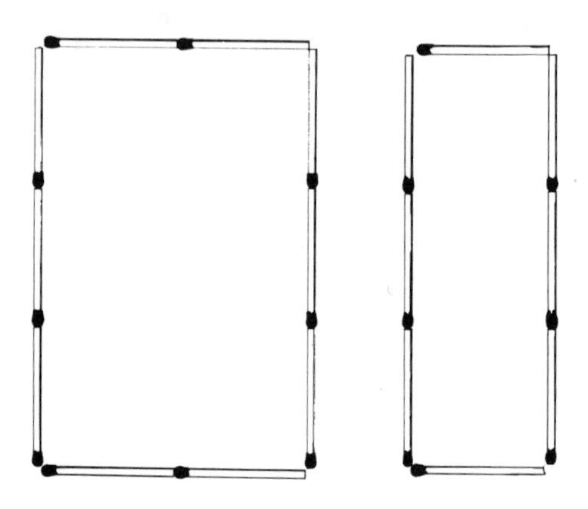

64. Put a match into the hollow between the thumb and forefinger of each hand as shown on the left. The trick: transfer both matches at the same time to the opposite hand, so they rest on the tips of the thumb and forefinger as shown on the right.

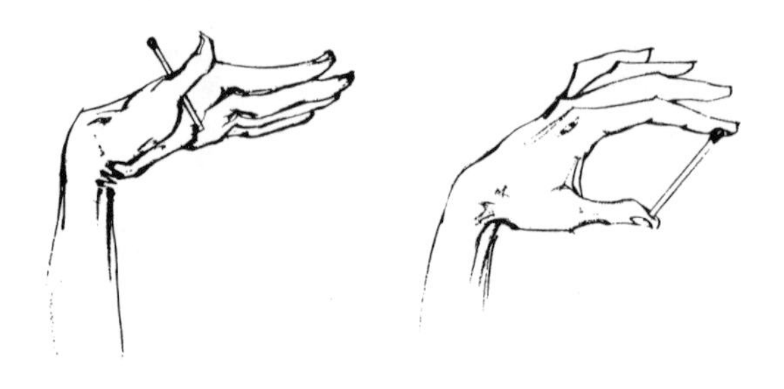

65. With six matches, form four equilateral triangles.

66. From this row of six matches, shift two so as to leave nothing.

67. Lay down nine matches on the table and lift them all at the same time with one more match.

The next seven problems are based on this two-by-two grid:

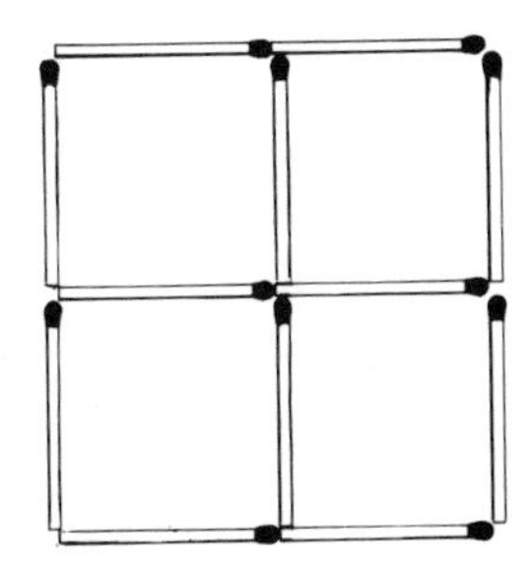

68. Move two matches and make seven squares.

69. Move four matches and make 10 squares.

70. Remove one match and move four matches to make 11 squares.

71. Remove two matches and leave two squares.

72. Remove three matches, shift two matches and form three squares.

73. Shift three matches and leave three squares.

74. Shift four matches and leave three squares.

75. Start with a pile of 11 matches. From these, remove five. Can you add four, and have nine left?

76. Make 11 squares with 15 matches.

77. A friend tears a paper match from a pack and makes you this proposition: "I'll toss this match into the air. If it falls on either of the broader sides, I'll pay for the drinks. If it falls on one of the narrower edges, you will pay for them." Is this a good bet?

78. With eight matches, prove that half of 12 is seven.

79. There's no trick to this one. It's simply an exercise in manual dexterity. Lay five matches on the table. Pick up one between your two thumbs (Fig. 1). Then pick up one between your forefingers; then between your second fingers; then between your ring fingers; and last between your little fingers (Fig. 2).

Now, lay the matches down, one at a time.

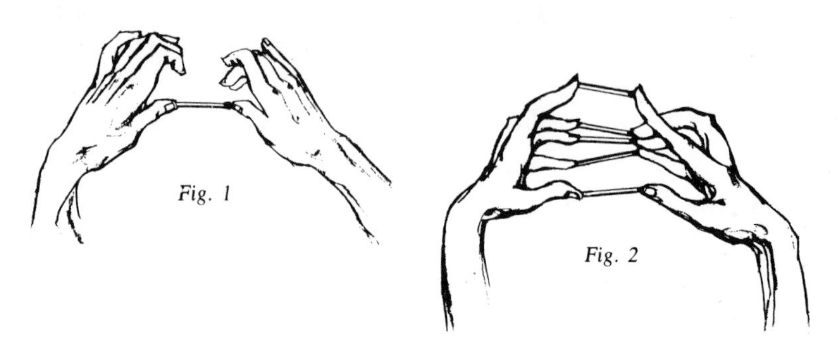

Fig. 1

Fig. 2

80. What is the smallest number of matches you need to do the following? Using matches all of equal length, form a square. Then on each side of the square form a right-angled triangle outside the square, so that each side of the square forms one side of a triangle, and no two of the four triangles are alike. (This is a hard one. The diagram gives you a clear idea of the problem, but does not necessarily show the proper sizes of the triangles.)

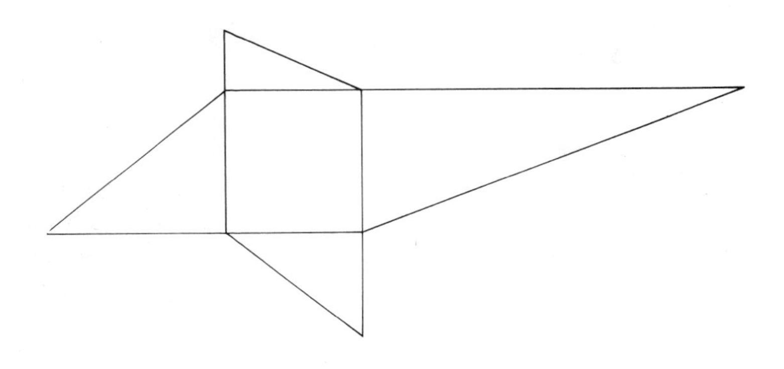

81. Three matches are put in a row with the center match reversed and its head pointing toward the player. The problem: pick up any two matches and reverse them, so that after three such reversals they are lined up with their heads pointing toward you. Although you do it again and again, others will find it impossible to make the moves correctly.

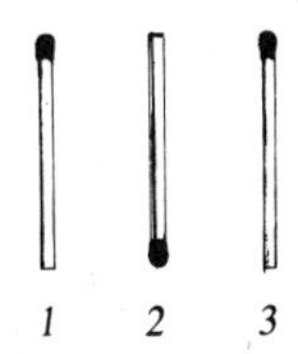

82. Remove four matches and leave nine squares.

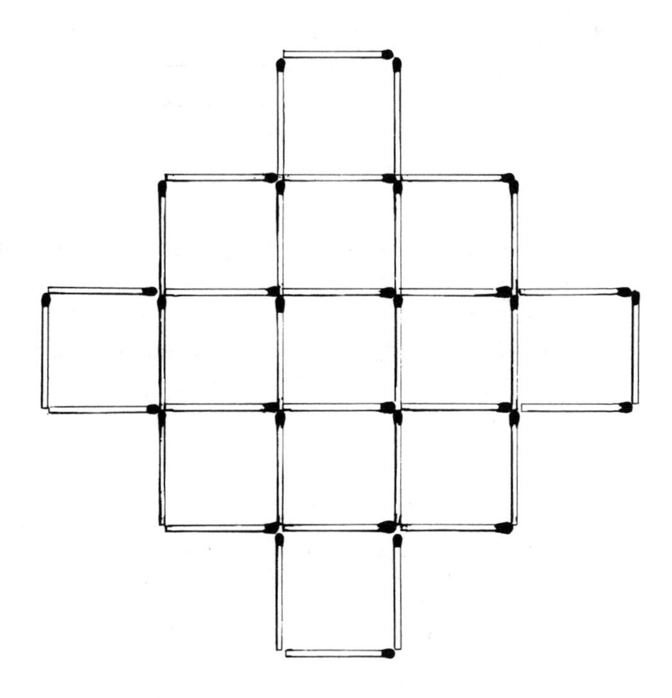

83. Bend four matches and place them thus. If you put a drop of water in the center (x), a four-pointed star will form.

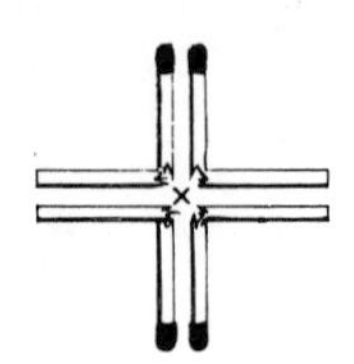

84. Move one match and make this equation valid.

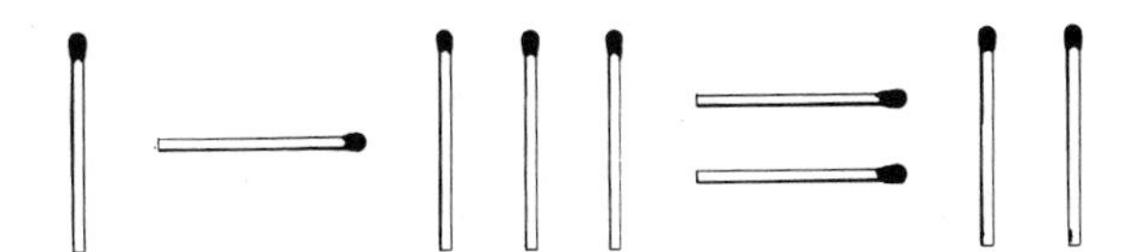

85. Place two matches upright in the end of an empty matchbox by sticking them between the cover and the drawer. Balance another match across the heads. Light the horizontal match in the middle and ask the spectators to guess which vertical match will catch fire first.

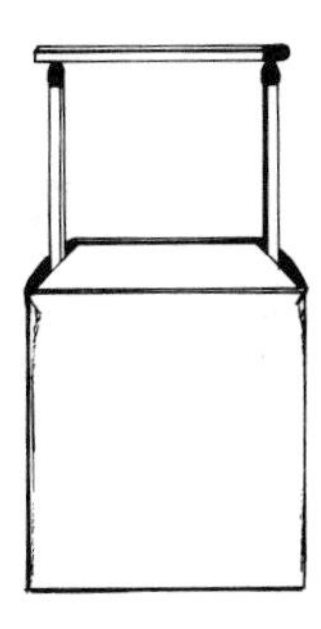

86. Move three matches and form four equal squares.

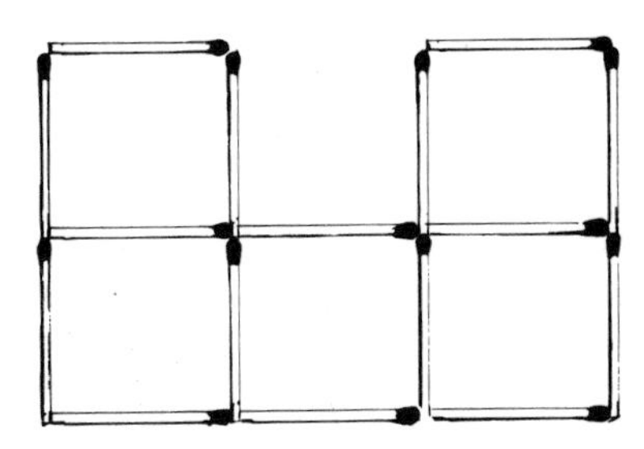

87. Can you hold a box of matches vertically about a foot above the table and drop it so that it falls on its end and remains upright?

88. Remove four matches and leave four triangles.

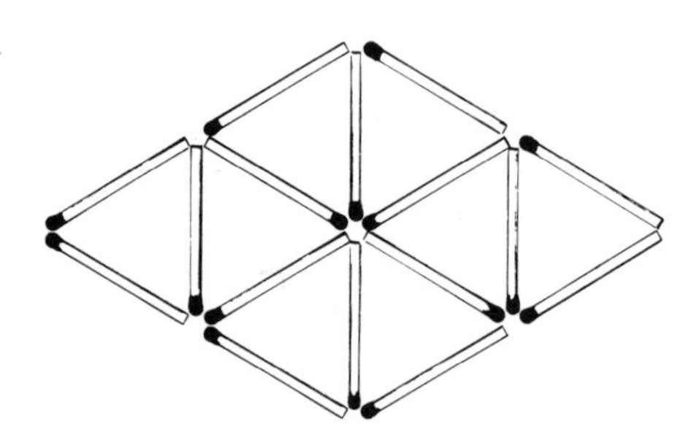

89. A matchbox is shown to be half-filled with matches; then it is closed and the box is turned upside down. The drawer is pulled out, still inverted, but the matches do not fall out. The box is shaken slightly and the matches are heard to rattle. The drawer is now pushed back into the cover and out the other side, whereupon the matches drop out immediately on the table.

90. Move one match and make the equation valid.

91. A match is laid across the open palm of the left hand (fingers pointing forward) so that the head projects over the right side. A second match is "charged" by rubbing it on the trouser leg and is brought up to the other match so that its head is just beneath the head of the match on your palm. As soon as the heads touch, the match in your palm flies several inches into the air.

92. Can you take 12 matches, each two inches long, and enclose an area of exactly 12 square inches?

93. Prepare a paper match by pulling it apart at the lower end and splitting it in half up to the head. Hold the match so that it looks unprepared, and light it. When the head burns past the break, slide the thumb and finger so that the two pieces fan out. It looks as though one match suddenly becomes two.

94. Here 13 matches form a triangular grid. Remove just three matches and leave three triangles.

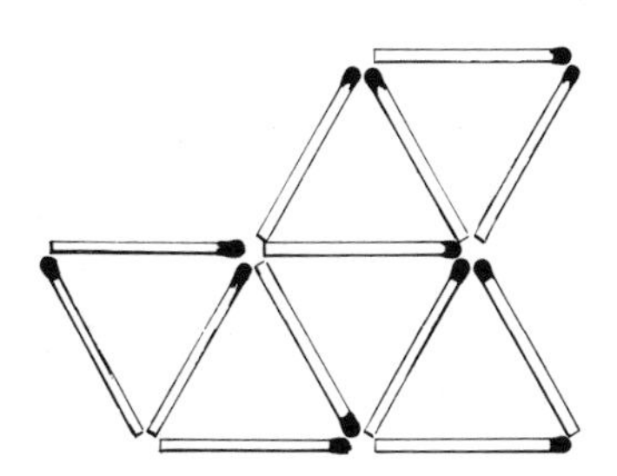

95. Make a small hole in the top of a matchbox near the end of the cover and push a match through it so that only the head remains outside. When you open the box by pushing the drawer forward, the match will push out of the hole into an upright position.

96. Move one match and make this equation valid.

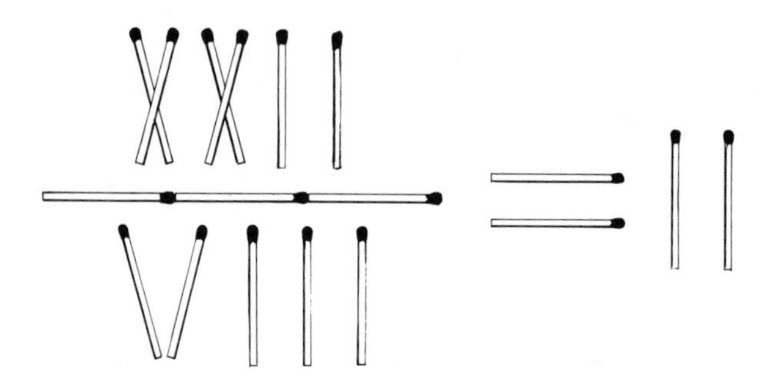

97. A Pythagorean triangle is a right triangle that can be formed with integral sides. For example, a 3, 4, 5 triangle is the simplest Pythagorean triangle.

What is the smallest number of matches needed to form simultaneously two different (non-congruent) Pythagorean triangles? The matches represent units of length and must not be broken or split in any way.

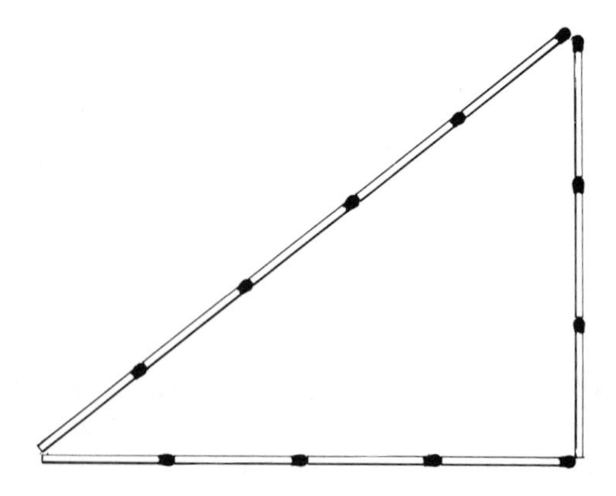

98. What is the value of *X*? (Hint: III can be interpreted in three different ways.)

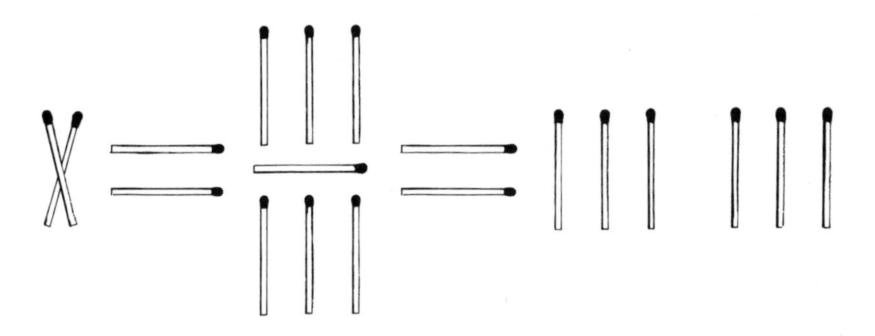

The ultimate in match-stick geometry has been reached by T.R. Dawson. In the *Mathematical Gazette* for 1939 (Vol. 23, pp. 161-8), he has shown that any geometrical construction that can be made with a compass and straight edge can also be made with a sufficient supply of matches. He first gives four postulates:

1. A match can be laid to pass through a given point or with either end at a given point.
2. A match can be laid to pass through two given points, or with one end at a given point and passing through the second given point. Two matches can be laid end to end but cannot overlap.
3. A match can be laid with one end on a given point and the other end on a given line.
4. Two matches can be laid simultaneously to form the sides of an isosceles triangle. Two of the ends coincide and the other two ends are at two given points.

Two constructions will be given.

I. To produce a given line: Repetition of the "half-hexagon" method will extend *AB* by successive units to any length desired.

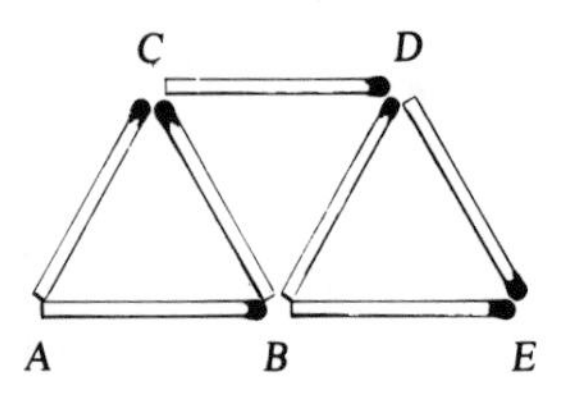

II. To bisect a given length less than that of a match: *A* and *B* are the two given points. Triangle *ACB* is formed with two matches. Triangles *AEC* and *BDC* are then formed. A match laid through *C* and *F* will bisect *AB* at *G*. This line also bisects angles *ACB* and *DCE*.

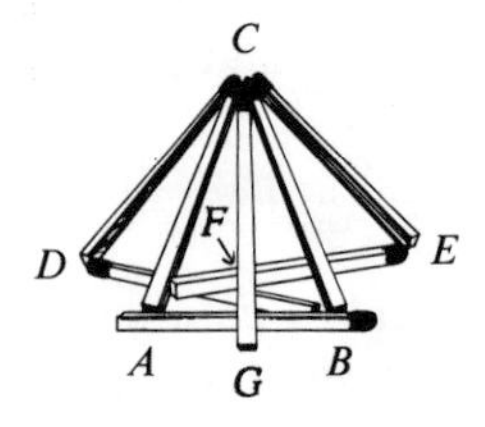

Now try your hand at these three constructions.

99. Bisect a length that is equal to that of a match.

100. Lay a parallel to a given line through a given point.

101. Construct a square, one match per side.

SOLUTIONS

1.

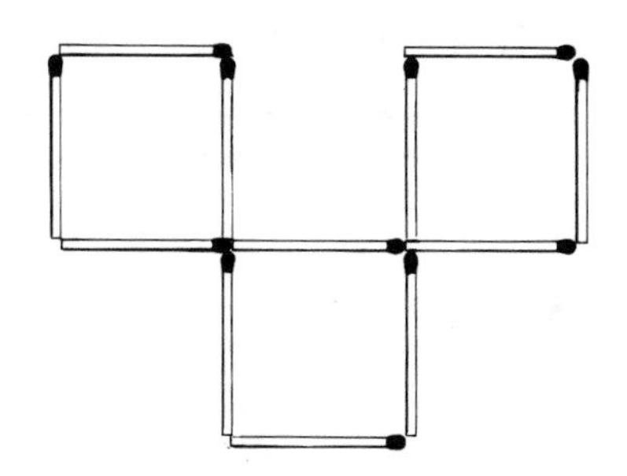

2.

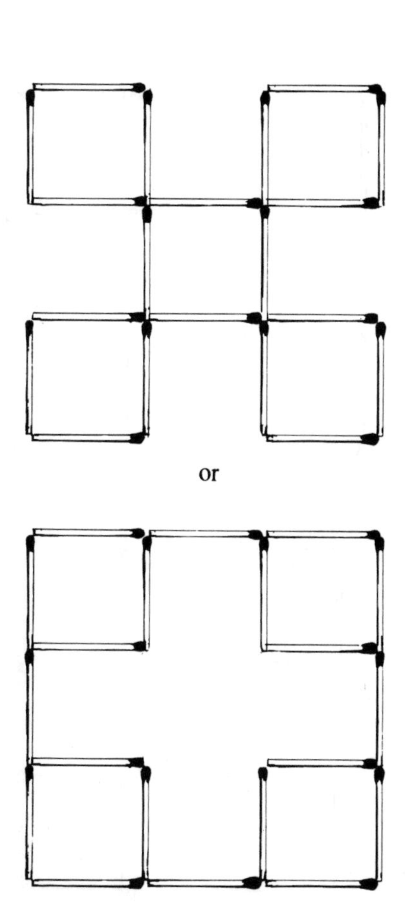

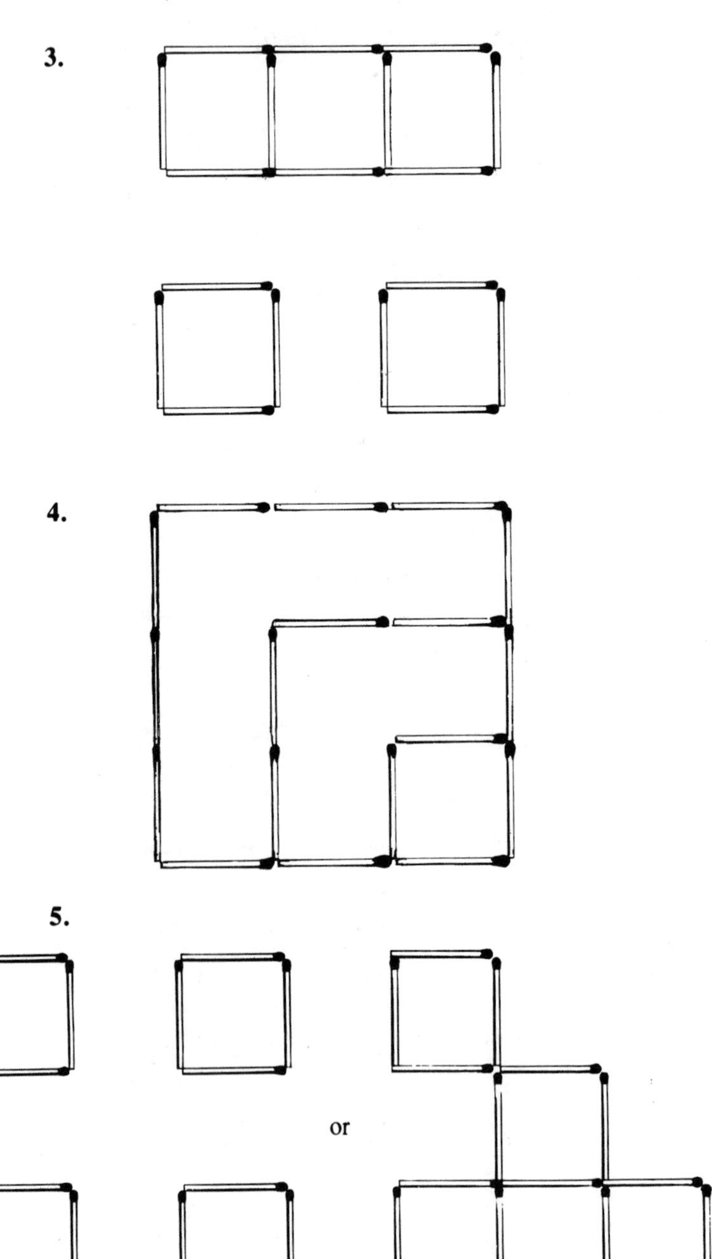
3.
4.
5.
or

6.

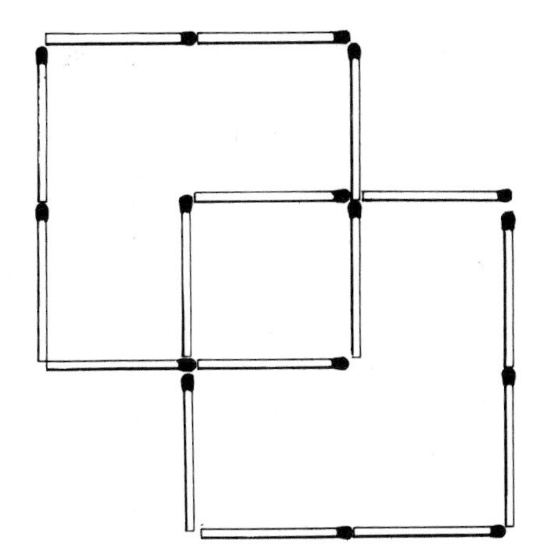

7.

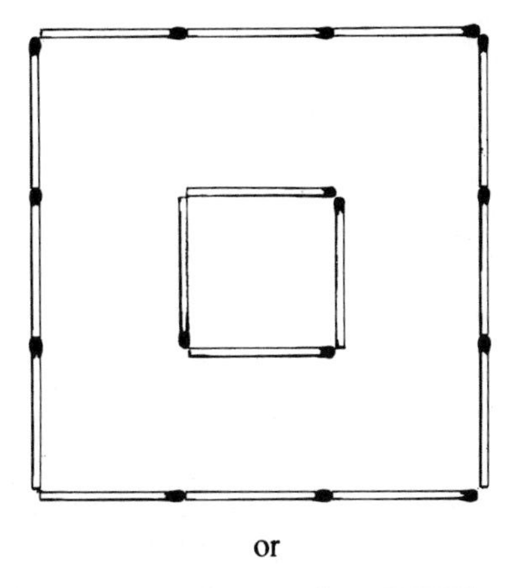

or

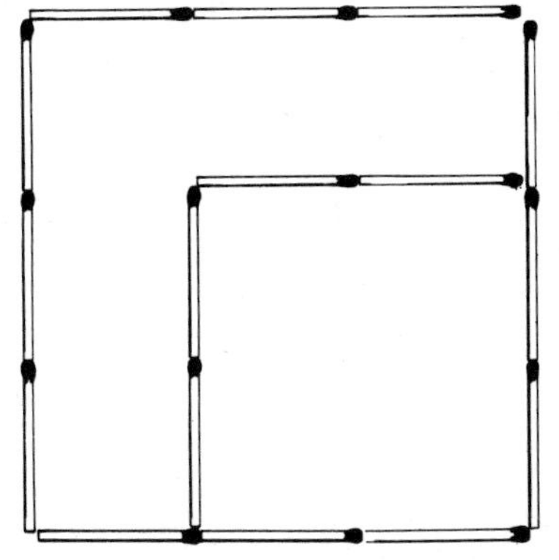

8. Four squares

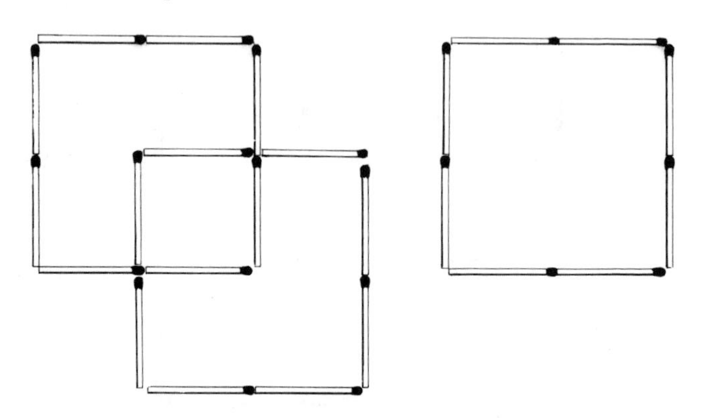

9. Five squares

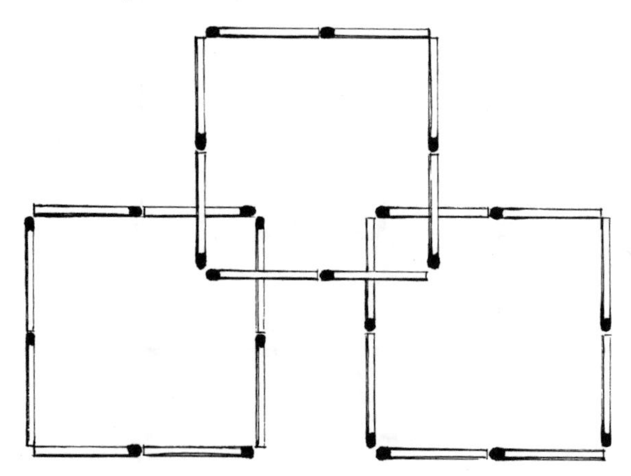

10. Six squares

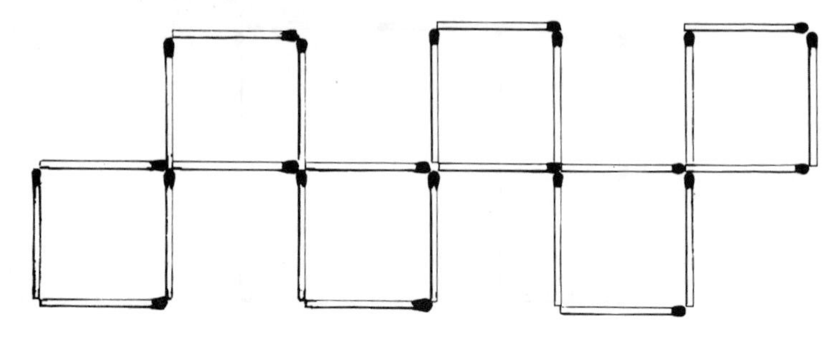

11. Seven squares

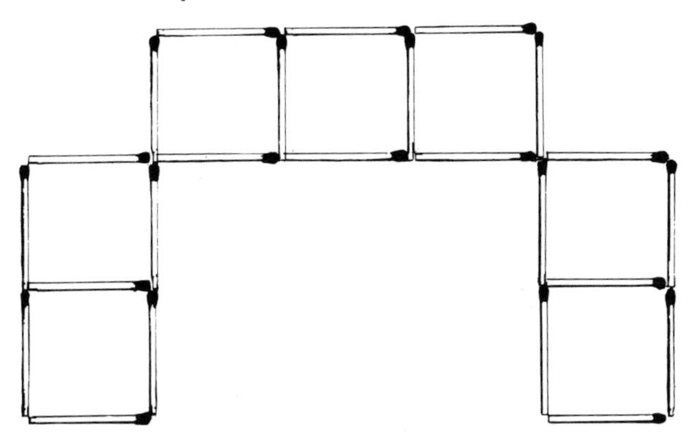

or

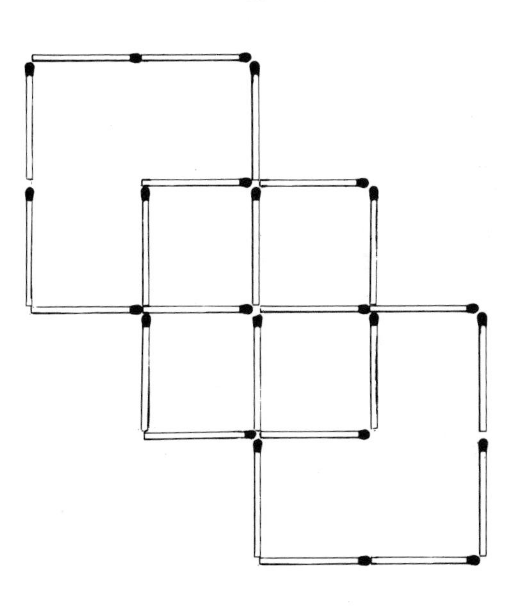

12. Eight squares

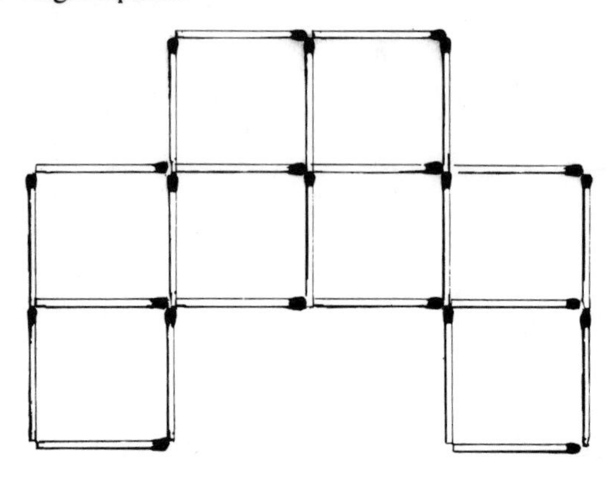

or

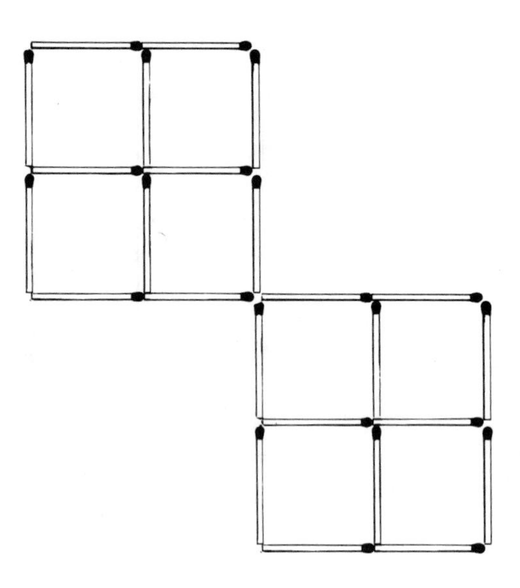

13. Nine squares

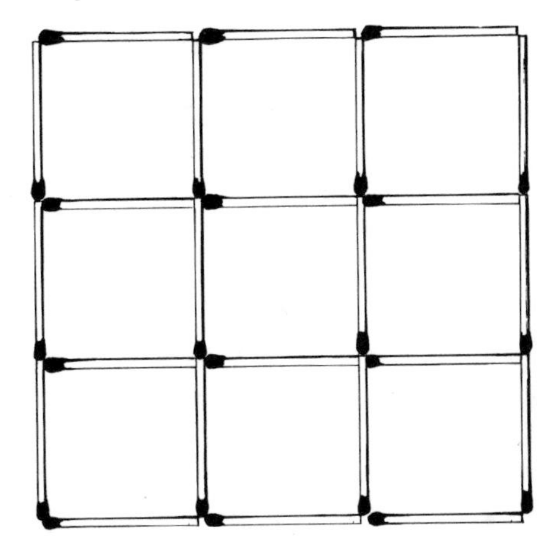

To carry this further, here is how to make 20, 42, and 110 squares from 24 matches.

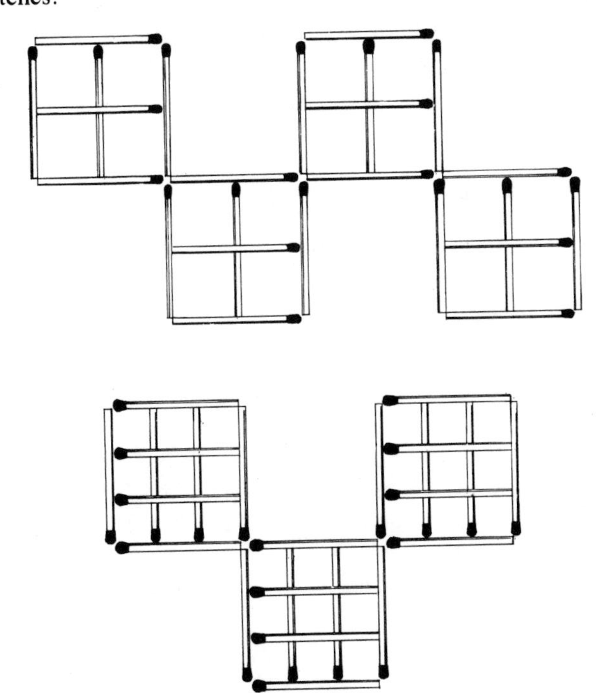

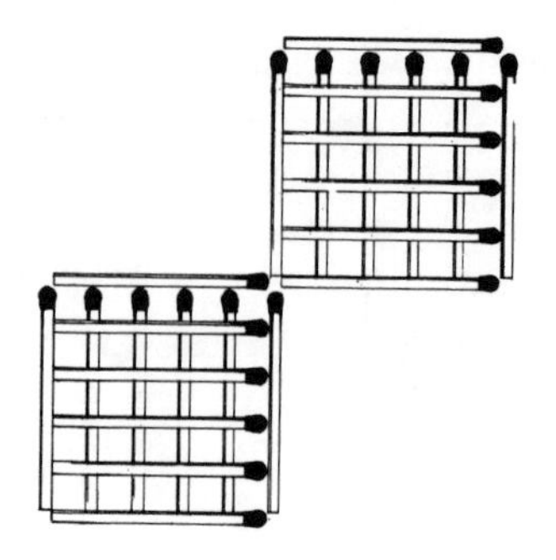

14.

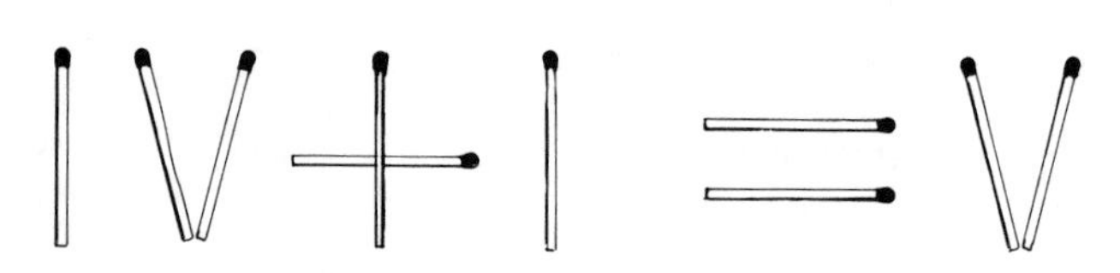

15.

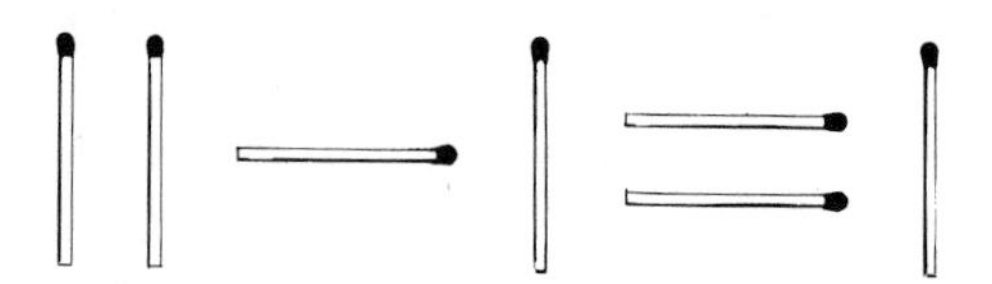

16.

1. 5 on 2		1. 4 on 7
2. 3 on 7	or	2. 6 on 2
3. 1 on 4		3. 1 on 3
4. 8 on 6		4. 8 on 5

17.

1.	5 on 1		1.	5 on 1
2.	6 on 1		2.	6 on 1
3.	9 on 3		3.	9 on 3
4.	10 on 3		4.	10 on 3
5.	8 on 14	or	5.	8 on 14
6.	7 on 14		6.	4 on 13
7.	4 on 2		7.	11 on 14
8.	11 on 2		8.	15 on 13
9.	13 on 15		9	7 on 2
10.	12 on 15		10.	12 on 2

18. You must maneuver your opponent into one of the following situations:

I. He has an even number of matches and there are 20, 17, nine, or four matches left in the pile.

II. He has an odd number of matches and there are 21, 16, 13, eight, or five matches left in the pile.

H.D. Grossman and D. Kramer *(American Mathematics Monthly* 52 [1945]: pp. 442–43) have devised a general strategy for the game. Parity is the oddness or evenness of the number of matches your opponent holds at any given time.

	I	II
If the maximum number of matches one may take is	even	odd
And if your ultimate object is to change your opponent's present parity, the winning move is to leave on the table a number of matches congruent to	1 (mod* $n + 2$)	1 or $n + 1$ (mod $2n + 2$)
And if your object is to keep your opponent's present parity unchanged, the winning move is to leave on the table a number of matches congruent to	0 or $n + 1$ (mod $n + 2$)	0 or $n + 2$ (mod $2n + 2$)

*The statement "a is congruent to b modulo d" is simply the mathematical way of saying "d divides a–b."

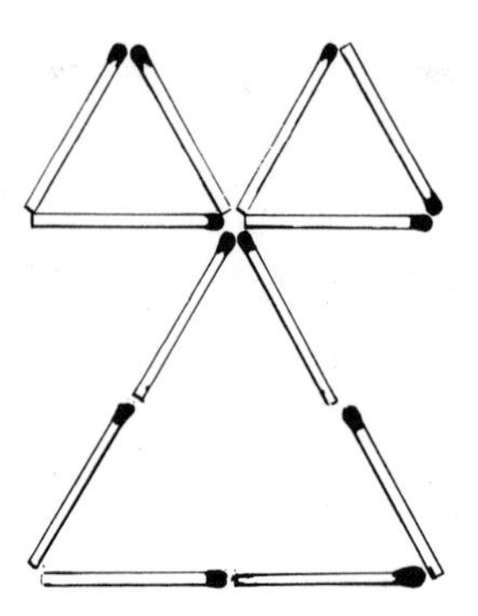

20.

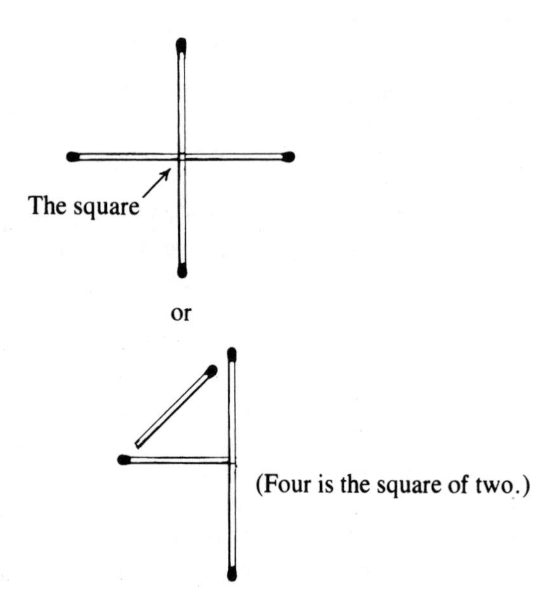

21.

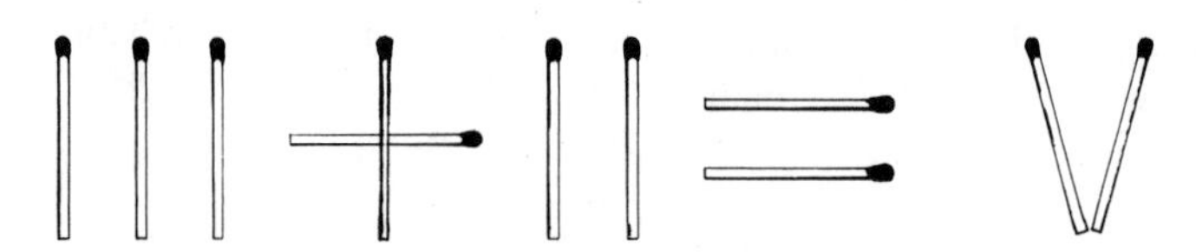

23.

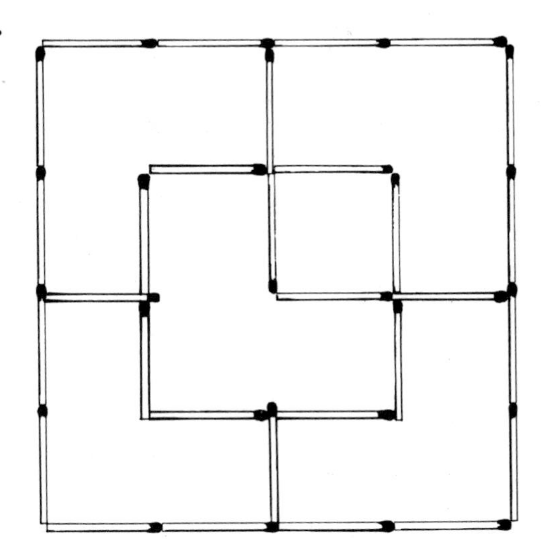

25.

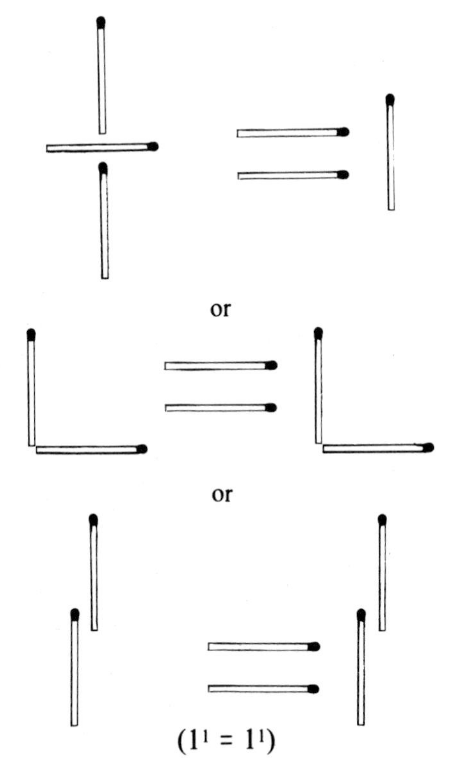

27.

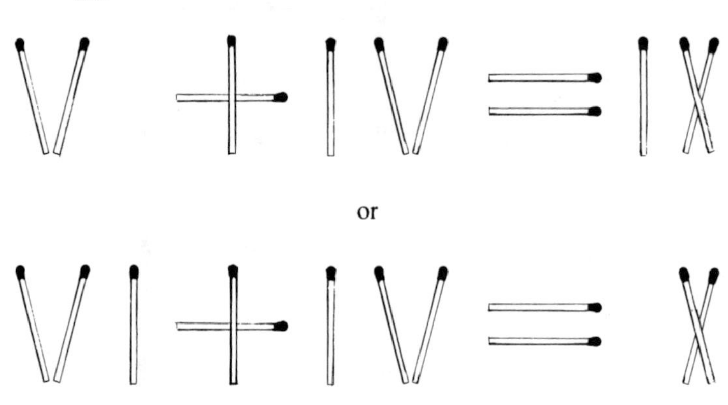

28.

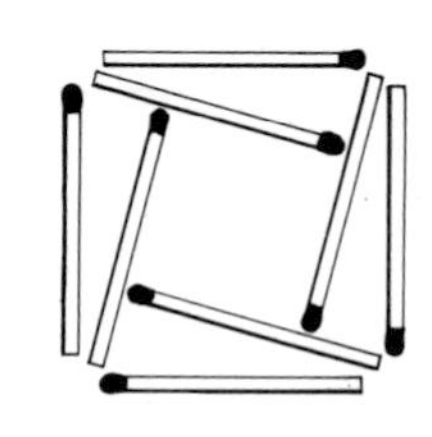

29.

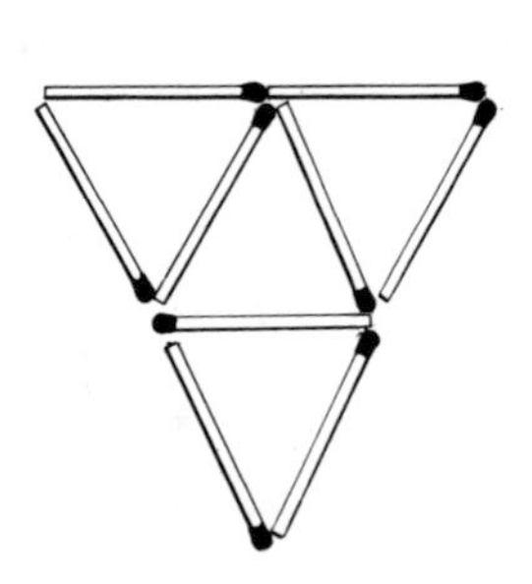

31.

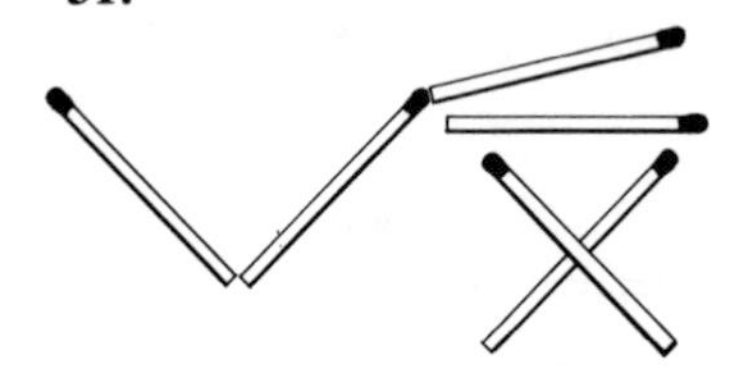

$$100 = \sqrt{\overline{10}} = \sqrt{10{,}000}$$

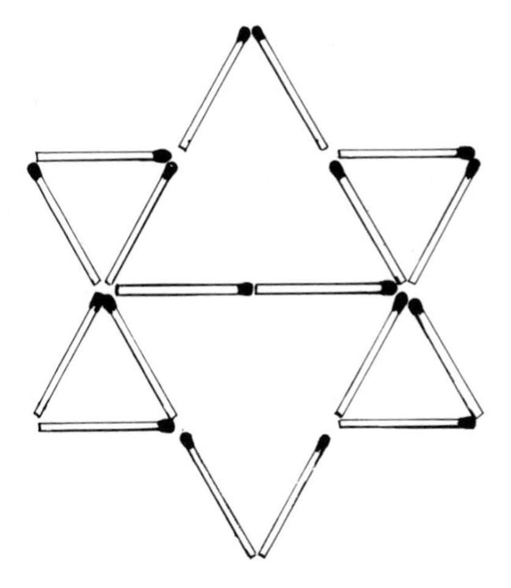

34.

In France there is a similar problem: Add three matches and make eight.

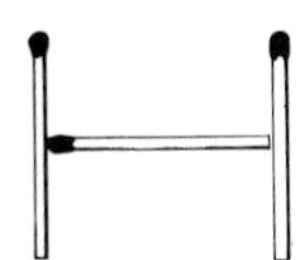
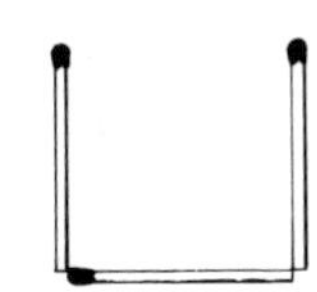
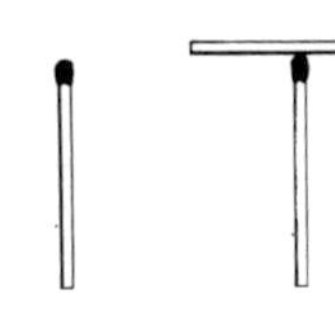

35. 420.

Squares	Rectangles	
1 x 1 — 35	1 x 2 — 58	2 x 7 — 4
2 x 2 — 24	1 x 3 — 46	3 x 4 — 22
3 x 3 — 15	1 x 4 — 34	3 x 5 — 14
4 x 4 — 8	1 x 5 — 22	3 x 6 — 6
5 x 5 — 3	1 x 6 — 10	3 x 7 — 3
	1 x 7 — 5	4 x 5 — 10
	2 x 3 — 38	4 x 6 — 4
	2 x 4 — 28	4 x 7 — 2
	2 x 5 — 18	5 x 6 — 2
	2 x 6 — 8	5 x 7 — 1

36.

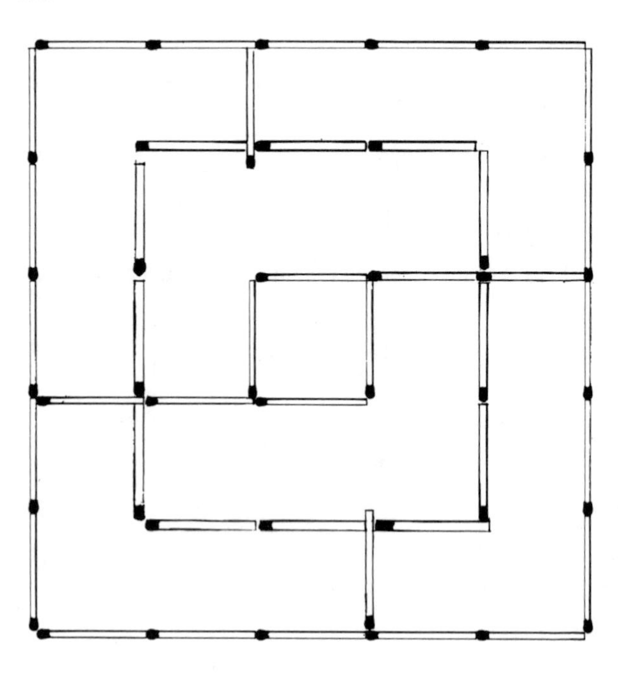

37.

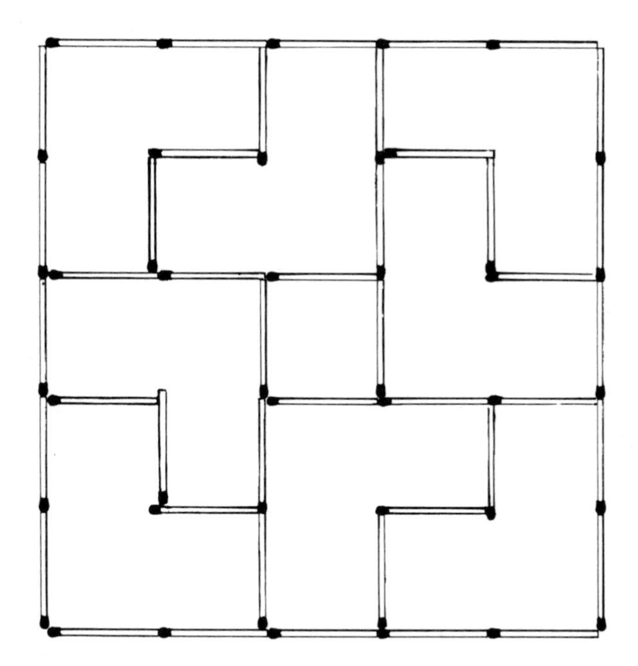

38.

39.

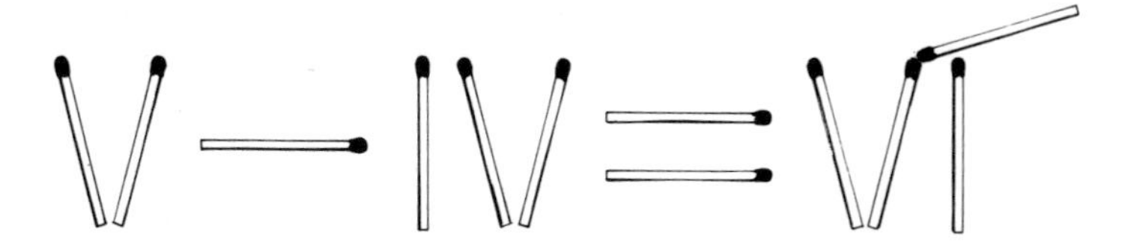

40.

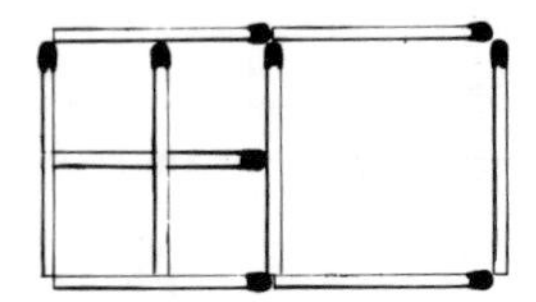

42. Simply walk around to the other side of the table. Or turn the book upside down.

43.

44.

45.

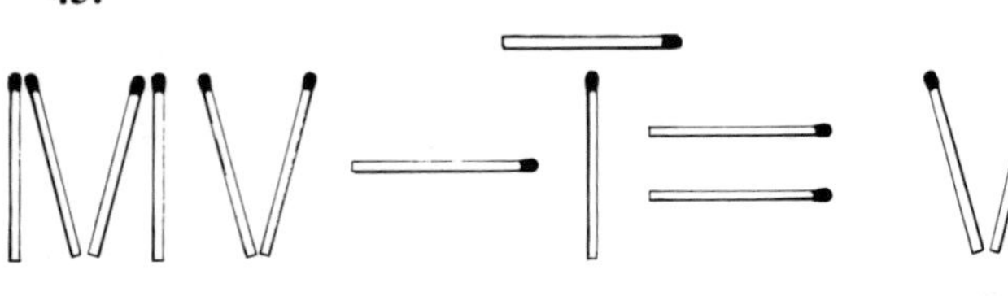

$\bar{\text{I}} = 1000$

46.

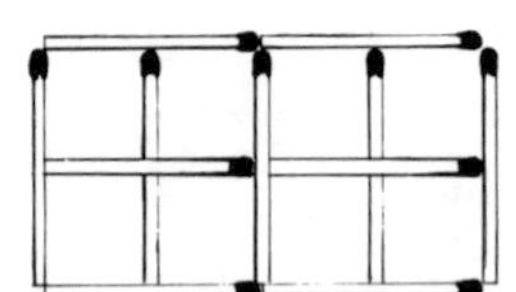

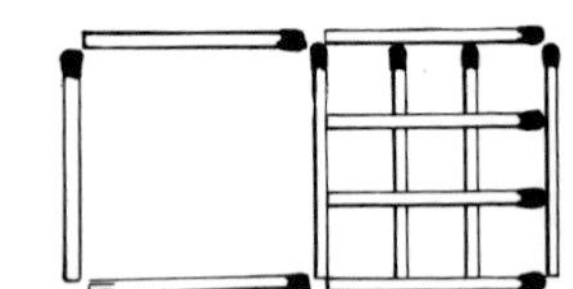

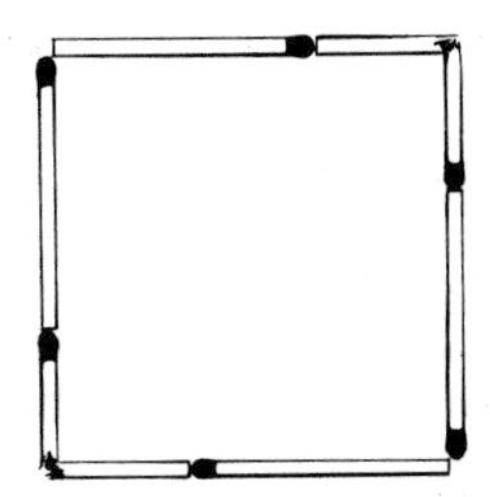

48. Light the match with another match and quickly blow it out. The head will stick to the glass so that you can pick the glass up and remove the coin. A regular match is better than a safety match.

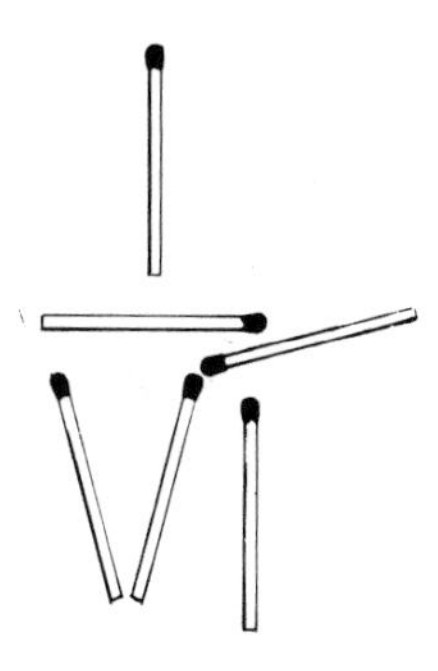

50.

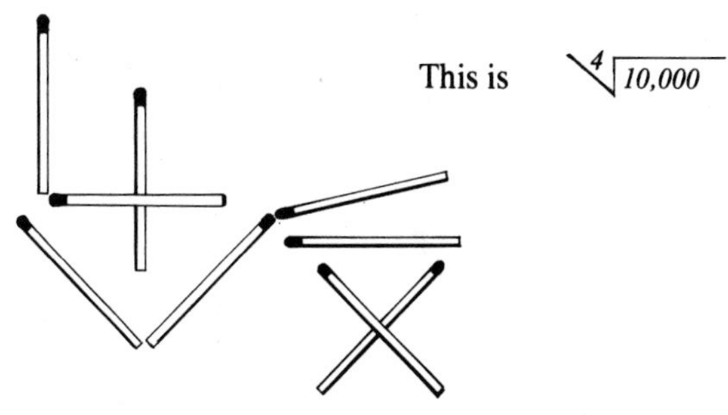

51.

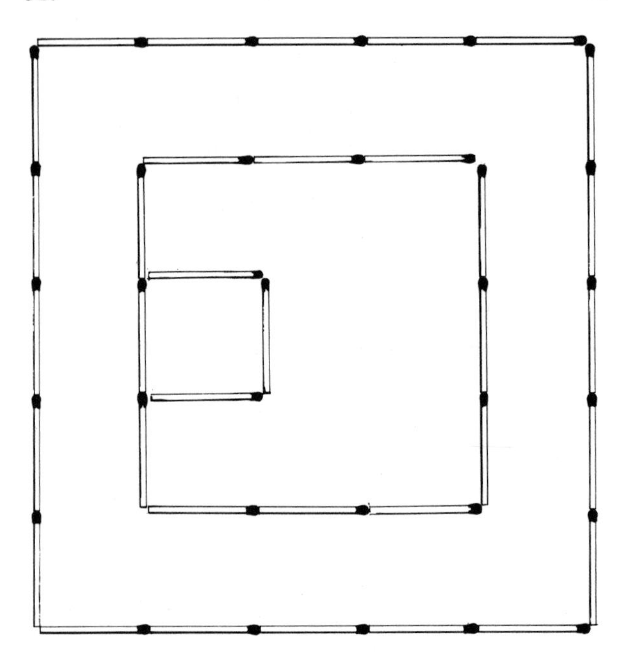

52.

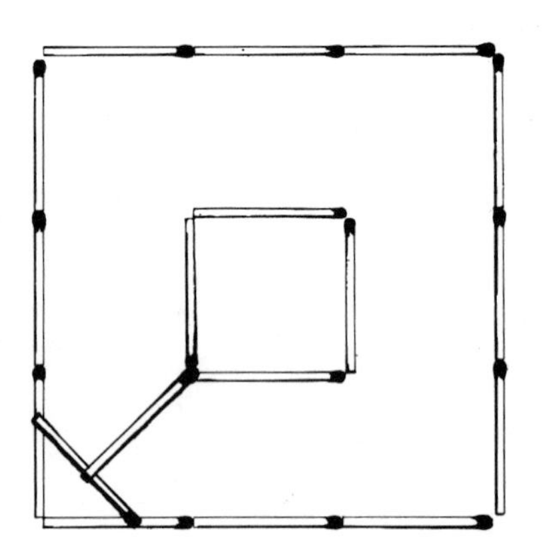

53.

54.

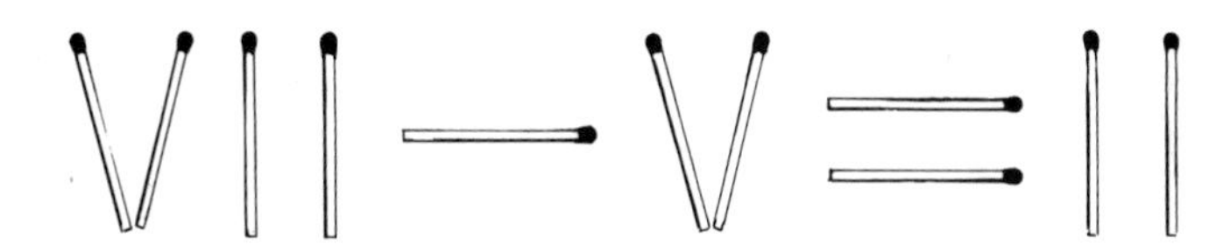

Also, of course,

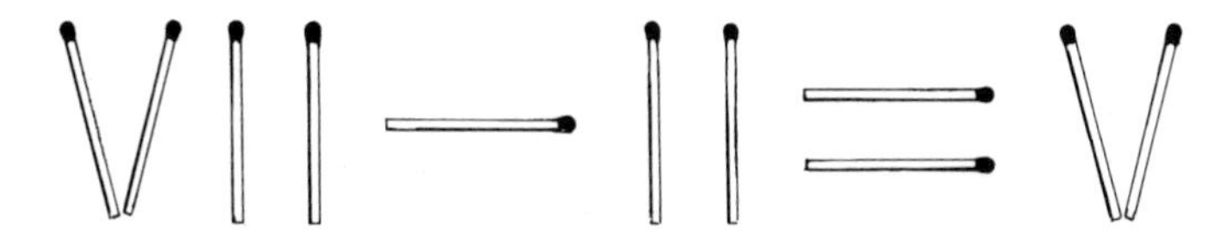

55.

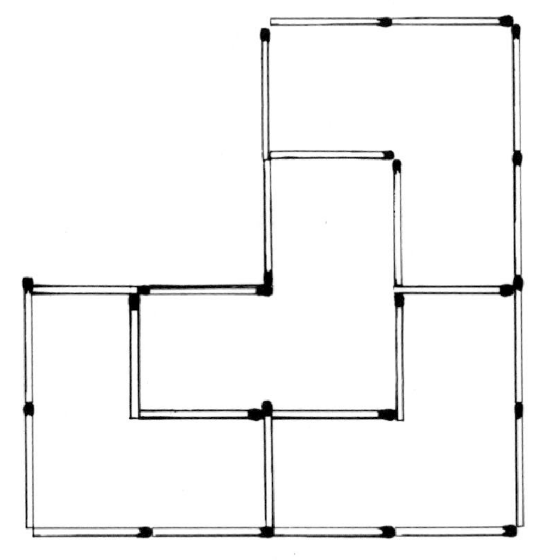

56. To do this, gently insert a fourth match horizontally. Tilt the two connected matches backwards until match *A* falls onto the one held in the hand. Then lower the horizontal match until the two connected ones fall down over match *A*. Then all three may be lifted.

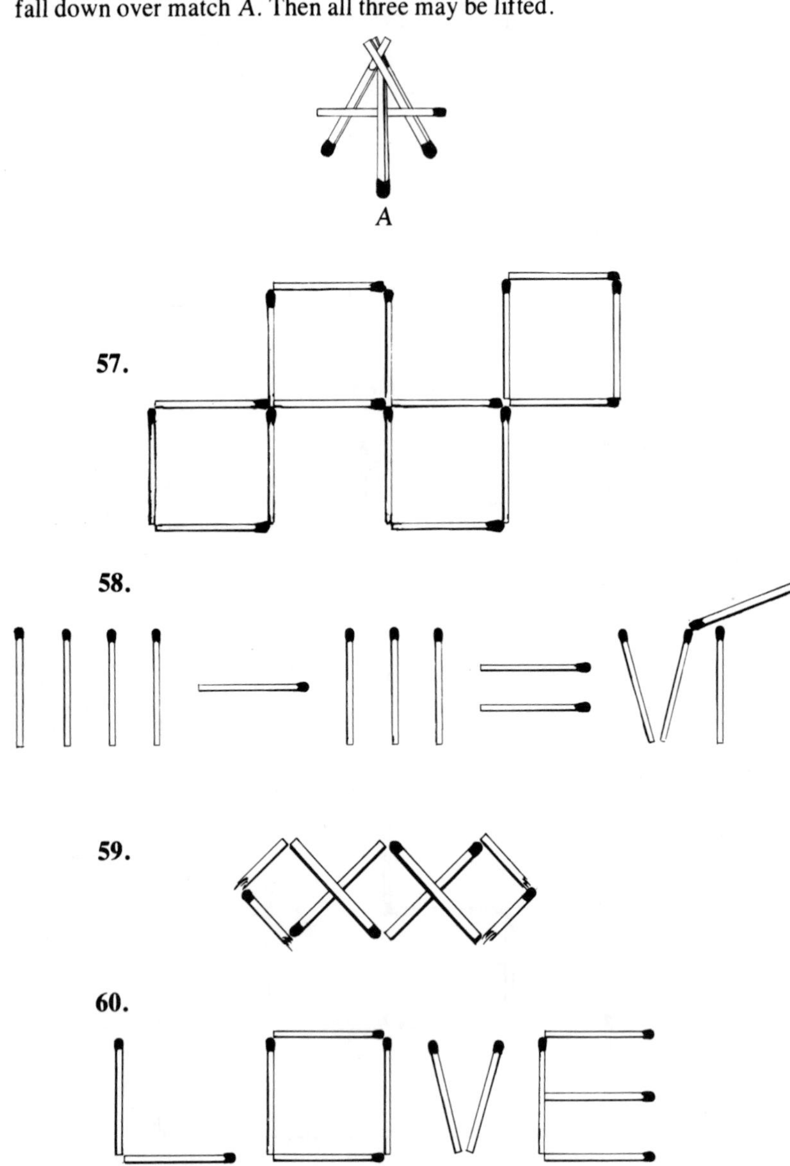

61.

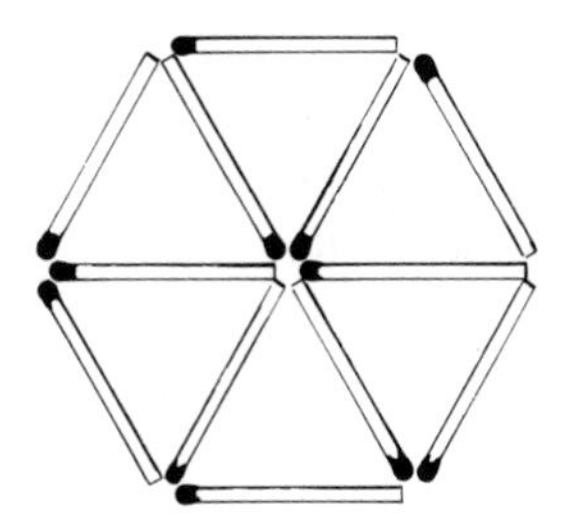

62.

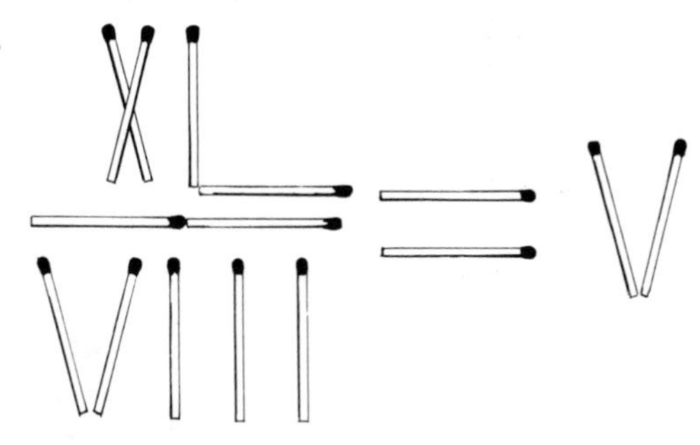

63.

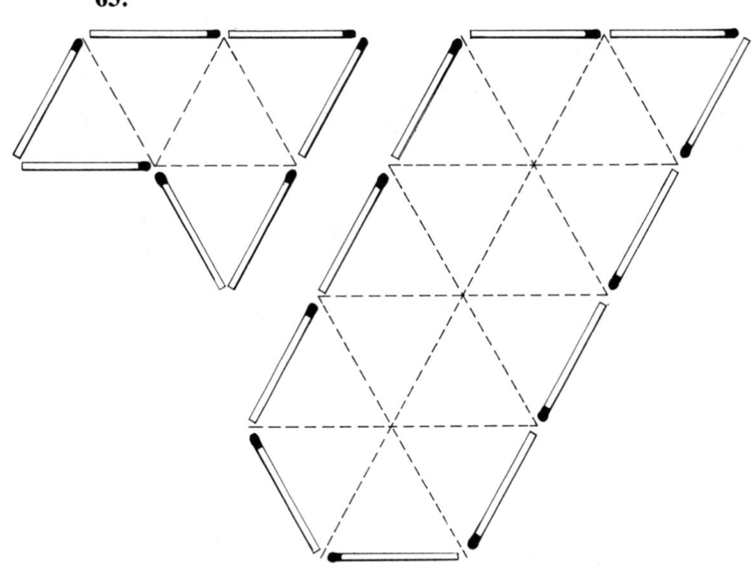

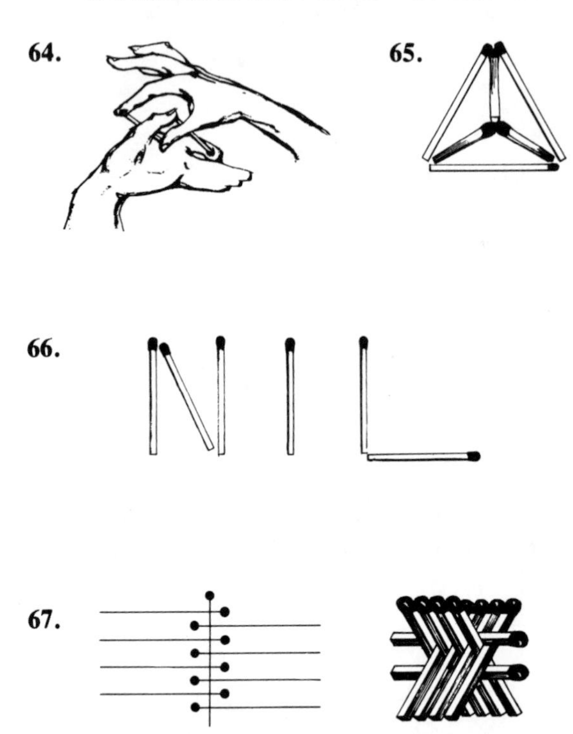

Lay the matches down as in the line drawing. Lay the 10th match on top of the others, parallel with the undermost match. When you pick up the first match laid down, the eight matches will fall into the form illustrated, and the weight of the top match will keep the others in position.

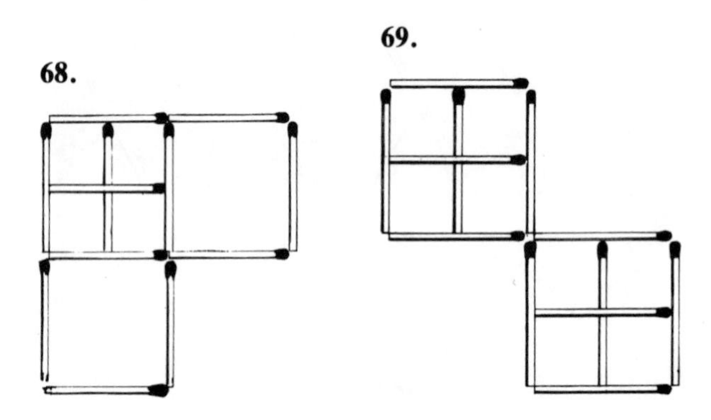

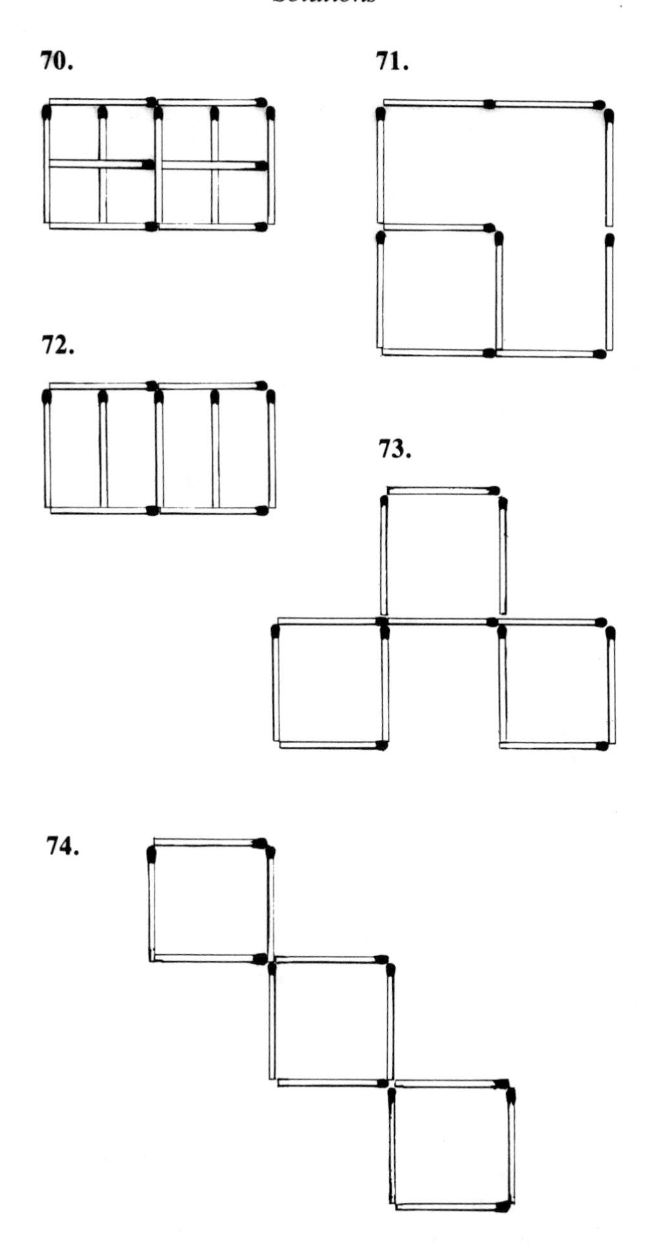

75. The four, of course, are to be added to the five *removed* from the first pile.

76.

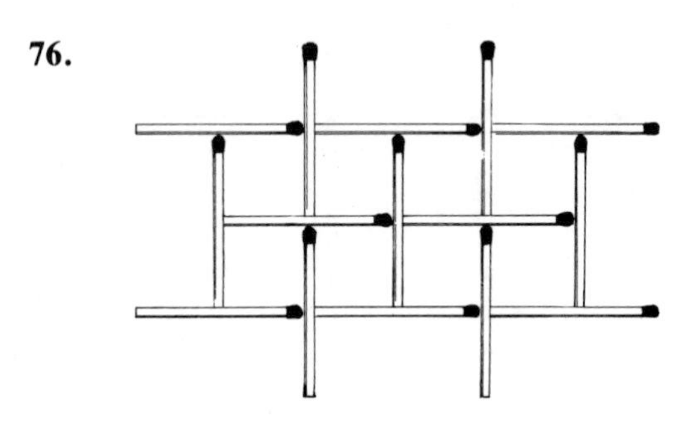

77. It is, for your friend. All he has to do is bend the match before he tosses it. It will be sure to fall on the edge.

78.

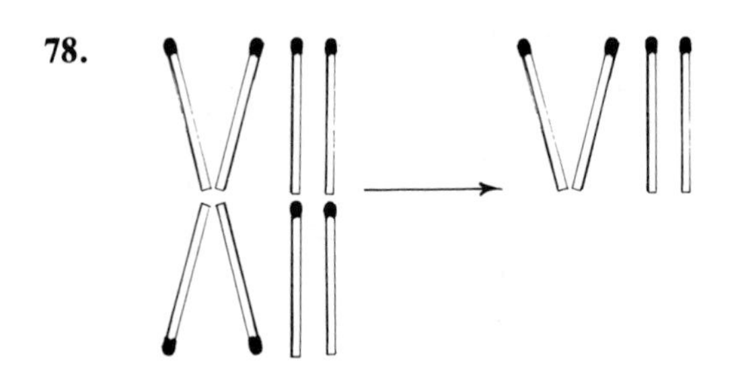

80.

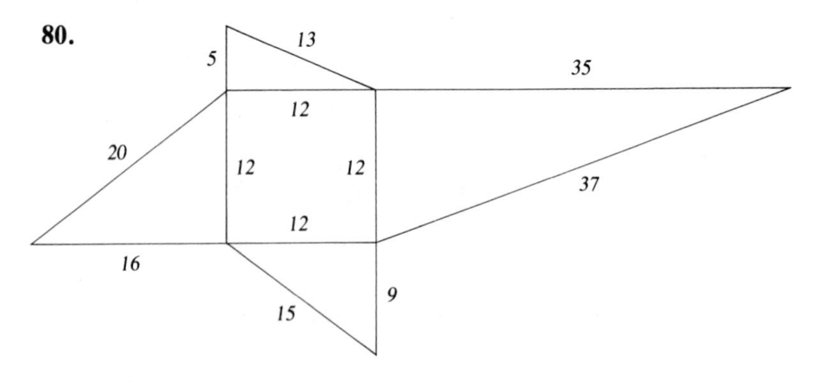

81. Here is one among several solutions: 1 & 2, 1 & 3, 1 & 2. The secret is to reverse the outer matches when you give the problem to someone else. In other words, the heads of the outer matches should point toward him and the center match should point away. No matter how many moves he makes, it will be impossible to do the trick.

82.

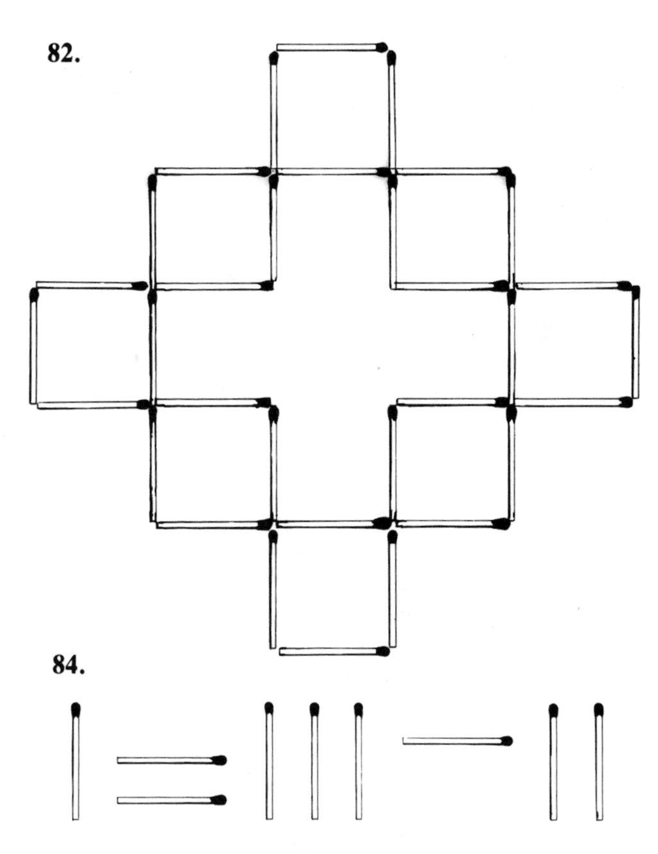

84.

85. Actually, neither will. The center match will burn a minute and then fall off.

86.

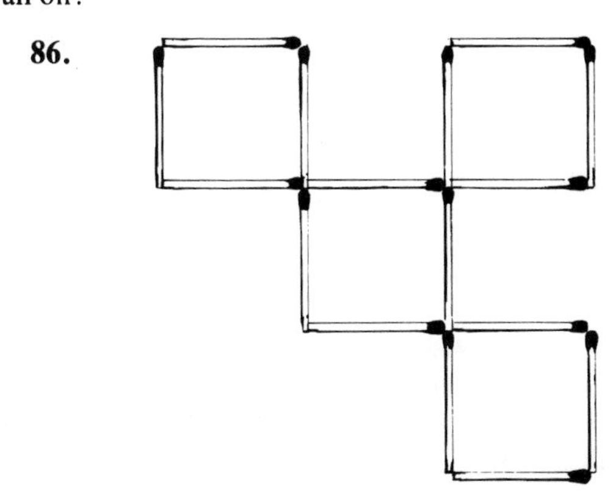

87. Before dropping the box, see that the drawer projects slightly at the upper end, but keep this concealed by your hand. When you drop the box, the force of the drawer sliding into place will cause the box to remain upright.

88.

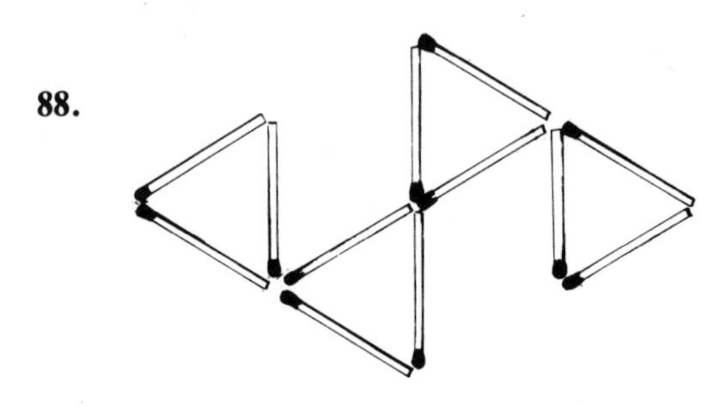

89. The trick is prepared beforehand by breaking a match to the proper length so that it can be wedged sideways in the center of the drawer. In removing the drawer, press the sides and hold the match in place. This prevents the matches from falling. When the drawer is pushed out the second time, your fingers do not press the sides, so the matches drop to the table and hide the smaller piece.

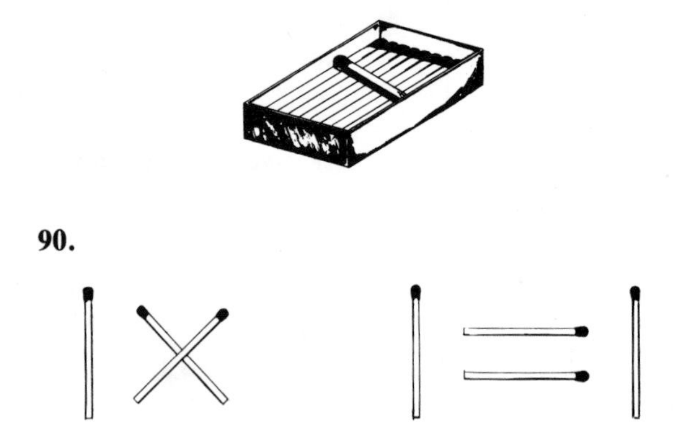

90.

91. The trick is accomplished by secretly snapping the lower end of the "charged" match. The snap is made with the fingernail of the second finger. The nail catches the lower end of the match, applies pressure upward, then slips off suddenly.

92.

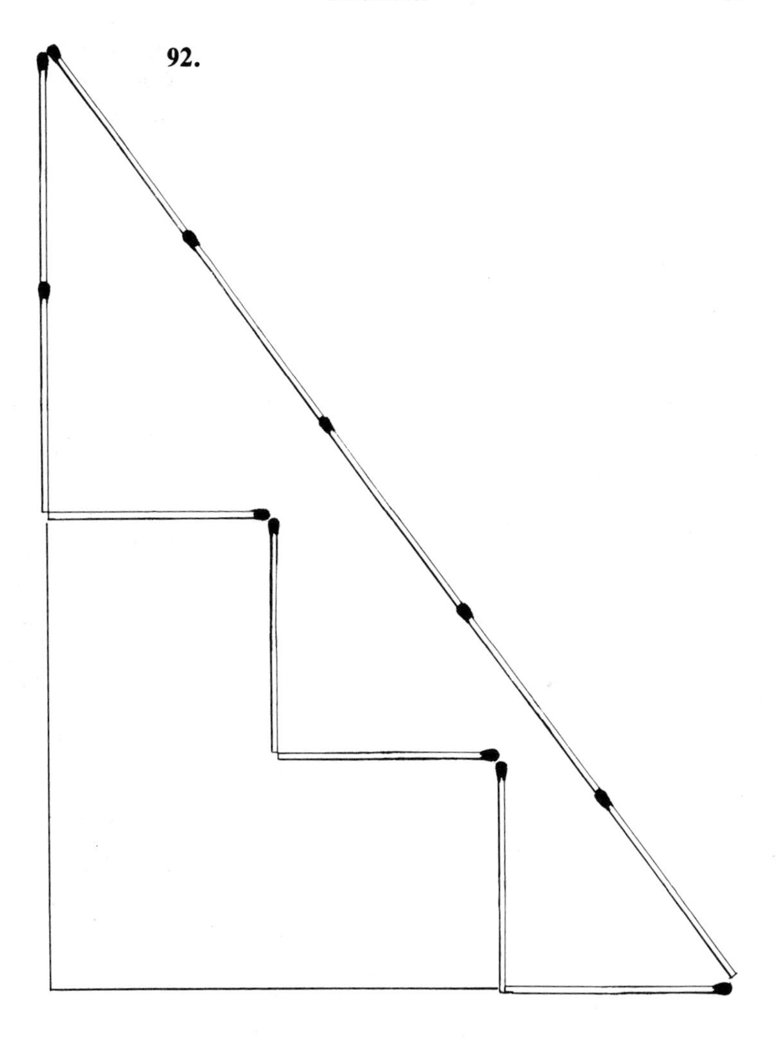

94.

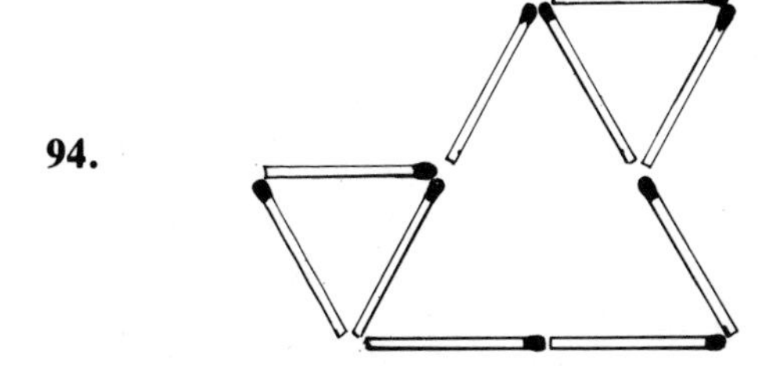

96.

97.

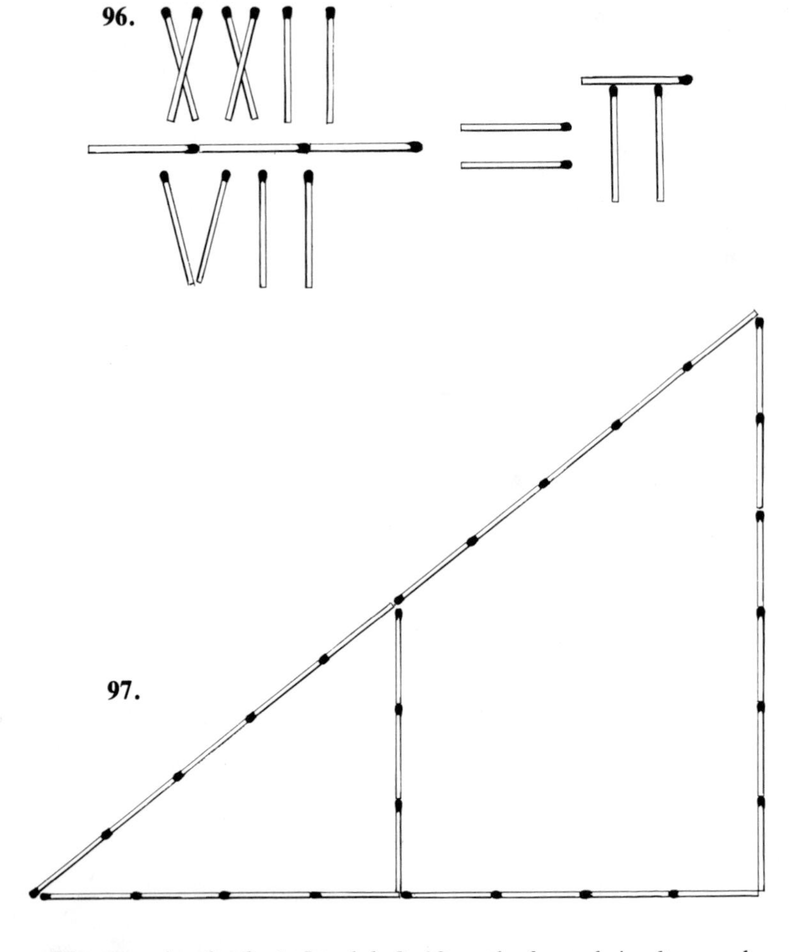

The two triangles 3, 4, 5 and 6, 8, 10 can be formed simultaneously on a plane with as few as 27 matches.

98. $X = \frac{111}{3} = 37$

In the fraction, the 111 above the line is in the decimal system. The 111 below the line is a roman numeral. The next 111 is also roman and the last 111 is in the binary system.

99.

AB is the distance to be bisected.
Triangle *BCD* is arbitrary.
Triangle *BDE* is constructed.
Triangle *BEF* is constructed.
Matches *AG* and *CG* are placed so their ends touch.
Triangle *AGH* is constructed.
A match laid from the intersection of *AG* and *EB* to the intersection of *AH* and *FB* bisects *AB*.

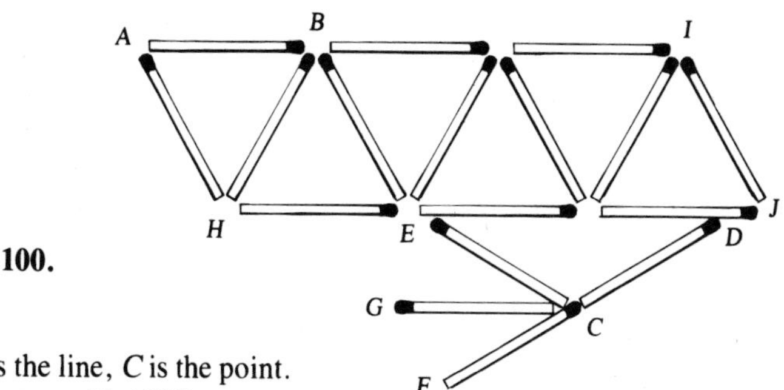

100.

AB is the line, *C* is the point.
Construct grid *AIJH*.
Lay matches *CE* and *CD* touching *HJ*.
Project line *DCF* by the half-hexagon method.
Bisect angle *ECF*.
CG is parallel to *AB*.

101.

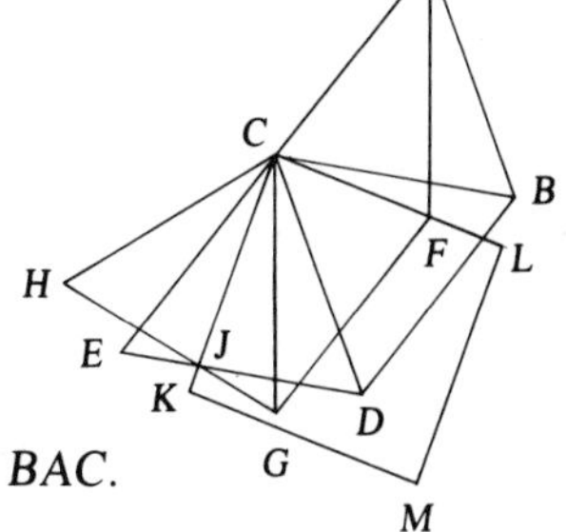

Construct grid *ABCDE*.
Lay match *AF* at random within triangle *BAC*.
Lay matches *CG* and *GF*.
Finish triangle *CGH*.
A match through *C* and *F* and one through *C* and *J* are at right angles.
Matches *KM* and *LM* complete the square.